AF846892

Springer Water

More information about this series at http://www.springer.com/series/13419

Marta Antonelli · Francesca Greco
Editors

The Water We Eat

Combining Virtual Water and Water Footprints

Springer

Editors
Marta Antonelli
University IUAV of Venice
Venice
Italy

Francesca Greco
Department of Geography
King's College London
London
UK

Springer Water
ISBN 978-3-319-16392-5 ISBN 978-3-319-16393-2 (eBook)
DOI 10.1007/978-3-319-16393-2

Library of Congress Control Number: 2015933824

Springer Cham Heidelberg New York Dordrecht London
© Springer International Publishing Switzerland 2015
This work is subject to copyright. All rights are reserved by the Publisher, whether the whole or part of the material is concerned, specifically the rights of translation, reprinting, reuse of illustrations, recitation, broadcasting, reproduction on microfilms or in any other physical way, and transmission or information storage and retrieval, electronic adaptation, computer software, or by similar or dissimilar methodology now known or hereafter developed.
The use of general descriptive names, registered names, trademarks, service marks, etc. in this publication does not imply, even in the absence of a specific statement, that such names are exempt from the relevant protective laws and regulations and therefore free for general use.
The publisher, the authors and the editors are safe to assume that the advice and information in this book are believed to be true and accurate at the date of publication. Neither the publisher nor the authors or the editors give a warranty, express or implied, with respect to the material contained herein or for any errors or omissions that may have been made.

Printed on acid-free paper

Springer International Publishing AG Switzerland is part of Springer Science+Business Media (www.springer.com)

Foreword

Food and Agriculture Organization of the United Nations (FAO)

The most recent world summit on sustainable development, held in Rio de Janeiro in June 2012 (Rio +20), clearly stated the central role of water in all human activities and in nature as a whole. Over the next 40 years, our planet will have to develop facing an unprecedented and increasing scarcity of natural resources, such as first and foremost water.

The main driver of this difficult future is the growing world population, which will quickly go from today's nearly seven billion people to over nine billion people in 2050; at that point, mere basic drinking and nutritional needs will put natural resources under extraordinary pressure, further exacerbated by the impact of climate change.

Food production will be the activity that will cause the biggest increase in water consumption, hence the FAO's slogan "The world is thirsty because it is hungry", coined for the 2012 World Water Day, dedicated precisely to water and food security. However, this close connection between water consumption (through crop's transpiration) and food production (through photosynthesis in plants) escapes most people. Many do not realize they may contribute to the reduction of water consumption through their eating habits. Lack of awareness of the damages and benefits to which different diets may lead is one of the main obstacles to sustainable development and to being respectful of future generations, from which we are borrowing extremely scarce natural resources such as water.

The Water We Eat comes at an important time in our journey towards gaining deeper knowledge about sustainable development. By means of the concept of "virtual water"—i.e., the water we consume to produce any kind of "product", and in particular to produce food—the book allows us to acquire and expand our knowledge of water issues, with the exciting potential outcome of significantly changing our eating behaviors. This is extremely relevant at the turn point of the post 2015 agenda for the achievements of what will be the *Global Development Goals*.

Moreover, we must note that *The Water We Eat* was published in 2013, the United Nation's International Year of Water Cooperation, which is another important event considering the fact that many nations will never be self-sufficient in the production of the food they need. Thus, satisfactory water (and subsequently food) security objectives can be achieved only through international cooperation.

Raising awareness about the fact that we can (and should) curb the "demand" for water by reducing food waste and following a balanced, low-water, diet is a pre-requisite to accomplishing sustainable development, and one to which all individuals and governments have the opportunity to contribute.

No better venues could be selected for presenting the translated version of this book at the Expo Milan 2015 'Feeding the Planet, Energy for Life'.

Pasquale Steduto
Former FAO Chief of the Water Unit, Chair of UN-Water 2007–1009
Presently Deputy Director Regional Office for Near
East and North Africa and Manager of the Regional Initiative
on Water Scarcity

Acknowledgments

The translation of the book was sponsored by Carlo Mazzola, President of Fondazione Eureka.

Fondazione Eureka was established in 2014 and is based in Milan; its objective is to develop and propose models of sustainable consumption of resources to reduce the impact that daily consumption habits have on the planet and, therefore, in our lives.

The Foundation develops research projects in partnership with prestigious Italian universities and research centers for rigorous and scientific analysis as well as numerical quantifications of environmental, economic, and social impacts of the consumption of resources, such as water.

The activities of the Foundation are addressed to the general public by providing user-friendly software applications to promote consciousness of citizens and to networks and study groups for the distribution of researches and the sharing of effective solutions.

www.fondazioneeureka.org

Contents

Part I
To Begin: Virtual Water and the Water Footprint

Not All Drops of Water Are the Same

Francesca Greco and Marta Antonelli

Where does the food we eat come from? Is it good for me? And am I doing the right thing in consuming it? In a world of limited resources, questioning ourselves about our lifestyles and our consumption patterns is not only desirable but also necessary. For this reason, we will introduce here the concept of "virtual water", namely the water required to produce the food, goods and services that we consume daily (Allan 1993, 1994, 1998). Thanks to the application of this concept, we will discover that we consume much more water than that we effectively see "running" before our eyes; we will highlight that the data showing Italian water consumption being 152 cubic litres a year per capita only reflects a partial consumption, referring only to the water used for our domestic purposes (drinking, cooking, washing, etc.).[1] The water we consume is actually much more than that. We are not able to perceive it as such because it is water that we literally "eat", embedded in an invisible way in the food we consume.

The purpose of this contribution is to unveil the consumption of virtual water, that is the volumes and the different typologies of water used in the production phases of goods and services of daily use. The study will focus on food products as they require far more water resources than any other good.[2] We will discover that

Both authors contributed equally to this work

[1]ISTAT data released for the World Water Day, 21 March 2011.
[2]2013 AQUASTAT data show the global percentages of use by sector: 70 % agriculture, 19 % industry and 11 % domestic use. Data available at: http://www.fao.org/nr/water/aquastat/water_use/index.stm

F. Greco (✉)
King's College London, London, UK
e-mail: francescagreco78@gmail.com

F. Greco
United Nations World Water Assessment Program (UN WWAP UNESCO), Perugia, Italy

M. Antonelli (✉)
IUAV University of Venice, Venice, Italy
e-mail: martaantonelli84@gmail.com

© Springer International Publishing Switzerland 2015

M. Antonelli and F. Greco (eds.), *The Water We Eat*,
Springer Water, DOI 10.1007/978-3-319-16393-2_1

the water used for the production of food follows quite alien and remote paths to our awareness as consumers.

1 Virtual Water, the Real Impact

The fact that water is a basic good and a human right, as proclaimed by the United Nations,[3] is now a part of our general thinking and accepted by all. Among the threats to water security in many countries of the world, we can include, above all, increases in population, climate change, economic development and the growth of consumerism, the general increase in the consumption of animal products and, finally, the asymmetrical availability of this precious resource for both economic and geographic reasons (see the map of physical and economic water scarcity in Fig. 2).

Water is a renewable resource but with very particular characteristics. Above all, it is a "primary good" essential for human and any other form of life. Water is an irreplaceable good, and its scarcity can be conditioned by physical or economic factors. Its scarcity, relative or absolute, is subject to natural processes which influence not only its geographical distribution, but also its access by humans in many parts of the world. Moreover, water is difficult to transport. It can be costly due to the use of expensive hydraulic works and can often be to the detriment of conserving and preserving natural environments and local populations. Even if it is a scarce and irreplaceable resource, water and its trade enjoy an extremely inflexible demand curve, and therefore, it is hardly influenced at all by market price increases (Savenijie and van der Zaag 2002). Furthermore, the exploitation of water by different production sectors involves significant environmental costs which, in many cases, lead to negative externalities for societies, not being "internalized" by those who produce them. This, in many cases, encourages the intensive use of the resource, well over its levels of sustainability, and consequently contributes to creating a very low efficiency in resource management.[4]

The awareness of our dependence on the ecosystems and the impact that our daily lives have on natural resources is, however, only partial. Indeed, most of us are ignorant of the fact that enormous volumes of water are involved in our daily

[3]On 28 July 2010, the United Nations General Assembly adopted Resolution A/64/L.63/Rev.1 stating the right to safe clean drinking water and to sanitation services as a human right essential to the full enjoyment of life and all other human rights.

[4]Different tools exist in the attempt to limit the externalization of environmental costs regarding the use of water for society. These include the following: sanctioning illegal water connections, setting up a pricing system for the distribution of water for irrigation and establishing a tax on groundwater pollution based on the principle "those who pollute pay". There is also a proposal for a tax on the use of agricultural land for building purposes that would place sanctions on the loss of agricultural land "sealed" by cement (see *Ambiente Italia 2012*, edited by D. Bianchi and G. Conti, *Edizioni Ambiente 2012*) (Bianchi and Conte 2012).

activities, often not actually being seen. In fact, the withdrawal and use of water by humans is not only limited to domestic use. Most of the water we use is the water we "eat", that is the water contained (even if we cannot see it) in the food that arrives on our table after having gone through the various phases of production, transformation and distribution (for more detailed information on the agri-food supply chain, see "Water and Food Security" by J.A. Allan in this book). In each of these three phases, water plays a fundamental role, both directly and indirectly, as a production input, destined, respectively, for a final or intermediate use (Antonelli et al. 2012). Some examples: a cup of coffee hides 140 liters of water, 135 in an egg and 2400 in a hamburger [data from the Water Footprint Network in Allan (2011), FAO (2012), Fig. 1]. As already mentioned, the water we consume is, therefore, much more than what we actually see. The virtual water content (measured in litres of water in the production of goods or services) is higher in food products, especially those of animal origin[5]. Therefore, the concept of virtual water is fundamental, not only in understanding our dependence on hydrologic systems, even those far away, but it also helps in understanding the impact of our lives, our daily activities and choices have on them.

2 Why Water is not All the Same

This contribution focuses on agriculture, as it is the major user of water resources at a global level (WWAP 2012). Contrary to our common understanding, the water that reaches our table in the form of food is not all the same. The different types of water involved in the production of agri-food goods can be divided into two categories—"blue" water and "green" water.[6]

"Blue" water is defined as surface water (found in rivers and lakes) or underground water (groundwater). It is easy to access and transport. It can be measured, stored in dams, conserved or pumped into water systems to meet the needs of different sectors (agricultural, industrial and domestic). Worldwide, according to FAO estimates (AQUA-STAT 2013), 70 % of this water is used in agriculture. In some countries, even very arid ones, the figure is even much higher than the world average, reaching more than 90 % of total water consumption. An example of this are the countries in the Middle East and North Africa, the most arid region in the world. Instead, "green" water is rain or snow water which falls to earth but does not become blue water (neither reaching the groundwater nor becoming a part of rivers, lakes or glaciers). This precipitation ends up evaporating or being transpired through plants. Its volume is equal to the volume of rainwater found in the body of the plant, to the water that produces natural soil moisture and that evaporates

[5]FAO 2010 AQUASTAT, http://www.fao.org/nr/water/aquastat/water_use/index.stm

[6]The theorizing of the concepts of blue and green water is the work of Malin Falkenmark (1989), a Swedish hydrologist and member of the Stockholm International Water Institute (SIWI).

Fig. 1 Virtual water in the more common foods. *Source* FAO 2012; edited by FAO WATER

naturally from plants during their life cycle. Green water cannot be transported nor withdrawn using pumps nor channelled. It is an intrinsic part of the plant–rain–soil system and cannot be appropriated from this.

To distinguish another type of water, we need to make a further breakdown in blue water: water originating from renewable or non-renewable sources. Surface water or groundwater belongs to the first group, being replenished by rainwater or snowmelt, where the exploitation threshold can be measured by calculating how much is naturally replenished by annual percolation. If the exploitation exceeds the natural replenishment threshold, we are speaking of the "non-sustainable exploitation" of a renewable source. Instead, the second type, non-renewable blue water refers to water extracted from groundwater, the so-called fossils, with a minimal replenishment percentage. This involves a stock of water that has been there for thousands of years and which, if consumed, will not be replaced for at least the same number of years. Even if not actually "non-renewable", this second type of water is considered to all effects as such because its total exploitation would result in certain water scarcity for hundreds of future generations (UNEP-DEWA 2003).

Returning to our earlier distinction between blue and green water, let us now explore how the latter has played an important role in global food production. Green water, that is rainwater evaporated from the ground during periods of crop growth, including evapotranspiration, allows for the growth of crops (non-irrigated agriculture) and the growth of vegetation and biodiversity conservation. While this type of water, differently to blue water, is completely invisible to the eye and relatively more complicated to measure, it accounts for about 84 % of the water used in agriculture (Fader et al. 2011) and its use has a less invasive impact on environmental balances (Aldaya et al. 2010). In an economic analysis, the opportunity cost of green water would be, moreover, very low, in some cases almost zero,[7] as it can only be used in agriculture and/or environmental conservation, and not in other sectors. Furthermore, its use does not affect the availability of blue water which, given it can be used in other sectors, has, instead, a high opportunity cost and needs to be protected as much as possible. In Italy, the yearly blue water volume per capita is 982 m^3, 61 % of total water availability, whereas green water amounts to 632 m^3 or 39 % of total availability (Gerten et al. 2011). As these figures show, green water is a very precious resource and plays a key role in water security and global food production (Aldaya et al. 2010; Allan 2011; Chatterton and Chatterton 1996). Therefore, the virtual water content of an agricultural product is the sum of the green water volume evaporated during the growth phase of a crop and the blue water volume withdrawn and used to grow the crop in a cultivated area. To this figure (the sum of green and blue water), we must also add the water needed to dilute the polluting agents during the production process, defined as "grey" water (Hoekstra et al. 2011). The different food products have a set virtual

[7]Different authors refer to green water as having a very low or zero cost opportunity. Some have suggested that green water could be considered as a "gift" (Chatterton et al. 2010; Chapagain and Orr 2009).

water content, generally expressed in litres or cubic metres, which can be broken down, in turn, into green water (from non-irrigated agriculture), blue water (irrigated agriculture) and grey water (polluted during the production phases). Therefore, a tomato irrigated with renewable water will have a lower environmental impact than one irrigated with non-renewable water. Moreover, the virtual water content of products of animal origin, such as eggs, milk and meat, is much higher than products of plant origin (Chatterton et al. 2010). However, the water sustainability of the production of a foodstuff will not derive only, as it would be easy to think, from the mere volume of the good's virtual water content[8] but, instead, will depend on the type of water used in its production (green or blue, renewable or non-renewable). Food produced by rainwater agriculture will have a lower impact compared to food produced using irrigated means, even more so in conditions of scarcity. This means that despite the fact that, on an average, 15,500 liters of water is needed to produce one kilo of beef, the meat produced by grazing livestock (non-irrigated) has a markedly lower water impact than meat from animals fed on fodder grown by irrigated methods (for more detail, see "The Water Footprint–a Tool to Compare our Consumption with the Use of Water" by A.Y. Hoekstra in this book).

In conclusion, not all drops of water are the same. The water found in everything we eat can have positive or negative effects on humans and the environment, in different countries, whether they be near or far from us, depending on the intrinsic characteristics of its original source.

3 The Origin of Virtual Water

Besides distinguishing between the different types of water contained in a product, another important step in understanding the water–food relationship is to identify the geographical origin of the virtual water contained in food products. In fact, the same product will have a different environmental impact depending on whether it is cultivated in a water-rich or water-scarce area. The surface areas of the earth can be subdivided into more or less wet zones, characterized by different climates and water availability—blue and green. The International Water Management Institute has divided the world zones into two macro-areas (Fig. 2): those where resources are plentiful (blue areas) and those where resources are scarce (orange, red and purple areas). It is interesting to note that water scarcity is not only assessed from a physical–natural viewpoint but also from an economic one (in cases where scarce economic means hinder the exploitation of naturally existing resources). Therefore,

[8]Note that the volume of water required to produce the same foodstuff can vary considerably depending on the production site—the productivity of water is, in fact, conditioned by the soil characteristics and climatic factors, by the technologies used and by the resource management methods. This means that the volume of water required to produce a tomato in temperate areas will be different from the virtual water contained in the same product coming from arid or semi-arid regions.

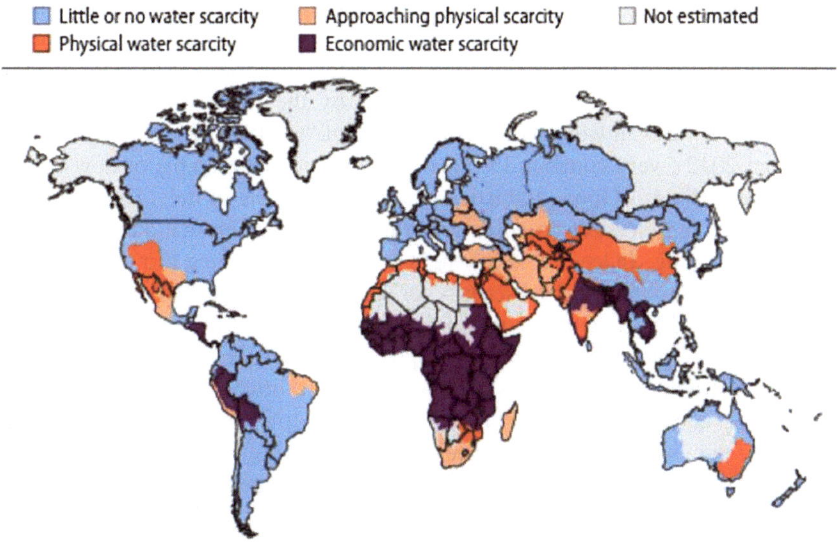

Fig. 2 Map of water scarcity. *Source* International Water Management Institute (2007)

water withdrawal and virtual water trade will weigh more heavily in local impact assessment depending on geographical origin. Looking at the map in the Figure, it is clear that exporting irrigated products from north-east America are quite different to exporting products from North Africa.

As far as the origin of virtual water from Italy is concerned, the production of agricultural products, both for internal consumption and export, presents a potential risk for local water resources in different areas, the so-called "hot spots". These are obviously found in the more arid areas of the country—in the south and on the islands—and involve south–north and south-external trade. From reading the hydro-geographical map of Italy, where most irrigation overlaps with the over-exploitation of groundwater and thus raises the risk of desertification, it is possible to identify the "risk zones", that is those zones where irrigated agricultural production results in a high environmental impact on the more critical water resources. These areas are found in Sicily, Apulia, Basilicata, Calabria and Sardinia. Moreover, in these areas, due to the intensive, and often unsustainable, use of groundwater (both renewable and non-renewable), the risk of desertification has increased. In Sicily, Sardinia and Calabria, the most exploited groundwater zones also coincide with the most intensely irrigated zones, while in Apulia we find in the most intensely irrigated zones less groundwater exploitation. This difference is due to the presence of a large canalization work, the *Acquedotto Pugliese*, using surface water for irrigation originating from outside the region.

Summing up, it is in those zones identified as "at risk" that goods are produced with a higher environmental impact on water.[9] If we add to these considerations those relevant to the social conditions of the migrant workers who work in these areas (FLAI-CGIL 2012), constantly reported by the Italian media, the general picture would be even worse. The phenomenon of "waste in the fields" (Segré and Falasconi 2012), very common in these areas, is another example of how controversial the methods of food production and distribution are in Italy. However, with the examples presented up to now, we do not want to claim that all virtual water trade is damaging. Indeed, it is a benefit where it contributes to relieving entire areas from the problem of food security, or simply, where it contributes to creating well-being for the locals and consumers without harming the environment. This occurs, for example, in cases where food products are exported by countries rich in water or in cases where products have a high green virtual water content. Virtual water trade becomes damaging in cases where it impoverishes the local water resources, the environment and the relevant populations, and therefore should be analysed concerning its full potential and all its benefits and implications. Considering that the biggest risks are linked to the negative impact of water use on the environment and on humans, this paper proposes to raise a point for reflection and bring to light some of the most hidden aspects of this phenomenon.

4 The Water Footprint of Italy

The concept of virtual water underlies the development of the so-called water footprint, an indicator for water consumption introduced by Arjen Hoekstra in 2003 similar to the "ecological footprint" developed in the mid-1990s by Mathis Wackernagel and William Rees. The water footprint of a person, a community or a company is defined as the total volume of water used to produce goods and services consumed by that person, community or company (Hoekstra et al. 2011). Water consumption is measured as the total water volume used and/or polluted in the production steps of a given good or service. The water footprint can be calculated for different types of subjects and groups of consumers (whether they be an individual, community, city or state) or producers (economic sectors or private companies). At a national level, we distinguish the water footprint of national consumption from the water footprint of national production. The former corresponds to the total of the internal water footprint (the water consumption in a given geographical area for a given period of time) and the external water footprint (water consumption originating from source external to the geographical area in consideration following the importation of virtual water involved in the international trade

[9]Further studies are being carried out to also identify all the cases involving over-exploited water used to irrigate products in Italy marked for exportation, both internally (south–north) and for foreign markets, at King's College London, London Water Research Group, Francesca Greco.

Fig. 3 Map of annual virtual water flows in Italy (1996–2005; mill. of m^3). *Source* Mekonnen and Hoekstra data, (2011); edited by authors

of goods and services). Instead, the national production water footprint is defined as the total of the internal water footprint and the water consumption for the production of goods for exportation (and, therefore, represents a water footprint external to the importing state).

Italy has a water footprint consumption per capita of 2330 m^3 per year, compared to an average of 1240 m^3, and it is the third net importing country of virtual water in the world after Japan and Mexico (Hoekstra and Mekonnen 2012, p. 3232–3237). Figure 3 shows the entry flows (import) and the exit flows (export) of virtual water for Italy linked to the international trade in agricultural goods, of animal and industrial origin.

What causes a high water footprint? There are four main factors that condition a country's virtual water footprint:

- water demand volume;
- demand composition which is mainly linked to the country's dietary habits. For example, as products of animal origin are more water demanding, the highest water footprint is found in the USA, followed by Greece, Italy and Spain;

- climatic conditions, which influence plant growth;
- agricultural practices, the management and efficiency of water use in the agricultural sector.

It is estimated that for every kilo of dry pasta produced in Italy, on an average, 1924 liters of water is required. The water footprint for a pizza weighing about 750 g is a little less at 1216 liters of water (Aldaya and Hoekstra 2010). It is important to underline that the environmental impact of the production of food goods such as these does not depend so much on the volume of water withdrawn and then incorporated in the final product, but rather on the context in which the water was withdrawn and used. The impact will be higher, for example in cases where the water withdrawn originates from over-exploited groundwater, as occurs in Sicily in the irrigation of durum wheat (*ibidem*). Instead, the cultivation of non-irrigated rice in Piedmont will have a much lower environmental impact (in fact, almost zero). The advantages arising from the introduction of water footprinting as an indicator of water consumption are varied. First of all, the water footprint is a recognition of the impact that human consumption has on the water resources by directly quantifying the volume of water incorporated in goods and services of common use. Secondly, the concept results in integrating a broader perspective into the traditional "basin vision" (interregional, interstate and international). The study of water footprints (internal and external) of different countries in the world has shown, for example, that many countries have, in fact, externalized their water footprint by importing from other countries those goods they need which require huge quantities of water in their production. This "movement" of virtual water between countries—as occurs in the trading of goods, especially agricultural ones—results, on the one hand, in satisfying the water–food needs of arid and semi-arid countries [first among these, the Middle East and North Africa, as shown by Allan (2001)], but on the other hand, gives rise, in some cases, to pressure being placed on the water resources of the exporting countries, in cases where they themselves can be exposed to a situation of water scarcity, both physically and economically.

4.1 Why Do We 'Eat' the Water of Other Countries?

The logic behind food production and trade is linked to political and economic reasons that are quite divorced from any considerations of an environmental nature. It is impossible to impose on any one country what they must or must not produce as the choice of agricultural policies lies solely under the control of that country. Therefore, the logic of production is quite external to the idea of maximizing the water resource use (for a study on water resource use in Italy, see "An economic approach to water scarcity" by A. Massarutto in this volume).

If the "good" use of water would fall under a moral, ethical code, a type of ecological conscience—similar to an animalistic conscience—then consumers choice could influence the market logic, which presently follows other codes.

Developing a "water conscience" could trigger a virtual circle of good practices on the part of consumers and companies that could make an investment and create an added value for products in the market (for how Italy could develop a national and global water conscience, see "Mobilisation for public water in Italy: a moral economy and virtual water" by E. Fantini in this volume).

A successful example of awareness raising on these issues has been provided by the experience of the WWF-Great Britain, which together with the Water Footprint Network, is working on drawing up international standards aimed at understanding how to improve the water footprint in the private sector of some of the largest global multinationals in order to decrease their products' footprint and create a guarantee for safeguarding water which would be recognized and included in their product brands (the so-called *water stewardship*). Water stewardship, that is the "safeguarding or good governance of water resources", includes the annual publication of a report on the water footprint of products, their labelling, certification of their business, comparison with other producers of similar products and reaching the quantitative targets in reducing their annual water footprint [10] (the opportunities and the challenges involved in introducing a sustainable water labelling for food products is the subject of a study "Aware eaters of water: an hypothesis for water labelling" by F. Greco and M. Antonelli (2012) in this volume).

5 Conclusions

This chapter has contributed to introducing concepts on virtual water and water footprint and has provided a preliminary analysis of the problems linked to the water we unconsciously consume through food with particular reference to the Italian case. We have seen how not all drops of water are the same since agriculture may use rainwater, with a very low or near zero environmental impact, or water originating from surface water bodies or from underground which is pumped and used for irrigation, whether it be renewable or non-renewable in nature. Consequently, despite appearances, not all tomatoes are the same. 70 % of all the world's fresh water is used in agriculture. The more negative implications of this occur when, for example, blue water is denied to poorer populations, in conditions of scarcity, or when non-renewable sources are used exceeding sustainability levels, in order to benefit the global market of food consumers.

The need to integrate the qualitative and quantitative (the volumes of water used) aspects has been recognized by the Water Footprint Network which has quantified, firstly, the water footprint of the different players (countries, companies, etc.). Opening up this debate in Italy, the third largest importer in the world of virtual

[10] Aldaya and Hoekstra (2010), "Analyzing International Virtual Water Trade and Water Footprint of Products" presented during the Corporate Water Footprinting and Managing Water Resources Meeting, London, 28–29 May.

water is therefore fundamental. Despite the fact that we are all informed about the origin of the water that runs from our taps and that this only makes up a small part of our total consumption, the information gap concerning the water contained in the food we eat, and which is the most important part of our needs, is still very general. This water often comes from far away with significant implications that we are not aware of (for further details on the issue of the de-socialization of water, see "Virtual water, H2O and the de-socialization of water. A brief anthropological overview" by M. Van Aken in this volume). Moreover, 90 % of the water used in food production is entirely managed by the private sector, specifically by a rather small number of multinationals (the so-called ABCD[11]) which operate in the international market and, thus, in conditions of "hegemony" (Sojamo et al. 2012). In conclusion, we believe that awareness is the underlying factor for change. Who would eat a strawberry knowing that its irrigation had denied drinking water to a village of Bedouins in the Arabian desert? If we could know, and choose, we would choose what is right.

Bibliography

Antonelli, M., & Greco, F. (2012). Una nuova visione dello spreco d'acqua. In A. Segrè, L. e Falasconi (Eds.), *Il libro blu dello spreco in Italia: l'acqua, Edizioni Ambiente* (pp. 129–168). Isola del Liri (FR).

Antonelli, M., Roson, R., Sartori, M. (2012). Systemic input-output computation of green and blue virtual water 'flows'. With an illustration for the mediterranean region. *Water Resources Management, 26*(14), 4133–4146

Aldaya, M. M., Allan, J. A., & Hoekstra, A. Y. (2010). Strategic importance of green water in international crop trade. *Ecological Economics, 69*, 887–894.

Aldaya, M. M., Hoekstra, A. Y. (2010). The water needed for Italians to eat pasta and pizza. *Agricultural Systems, 103*(6). doi:10.1016/j.agsy.2010.03.004.

Allan, J. A. (1993). Fortunately there are substitutes for water otherwise our hydro-political futures would be impossible. Priorities for water resources allocation and management, ODA, pp. 13–26.

Allan, J. A. (1994). Overall perspectives on countries and regions. In P. Rogers, P. Lydon (Eds.), *Water in the Arab world: Perspectives and prognoses* (pp. 65–100). Cambridge: Harvard University Press.

Allan, J. A. (1998). Global soil water: A long term solution for water-short Middle Eastern Economies. In *Proceeding of Water Workshop: Averting a Water Crisis in the Middle East—Make Water a Medium of Cooperation Rather than Conflict*, Green Cross International.

Allan, J. A. (2001). *The Middle East water question: Water and the global economy*. Londra: IB Tauris.

Allan, J. A. (2011). *Virtual water: Tackling the threat to our planet's most precious resource*. Londra: IB Tauris.

Bianchi, D. & Conte, G. (2012). *Ambiente Italia 2012. Acqua: bene comune, responsabilità di tutti*. Edizioni Ambiente: Milano, Italy

Chatterton, L., & Chatterton B. (1996). *Sustainable dryland farming: Combining farmer innovation and medic pasture in a Mediterranean climate*. Cambridge: Cambridge University Press.

[11]The reference is to the companies Archer Daniels Midland, Bunge, Cargill and Louis Dreyfus.

Chatterton, J., Hess, T., & Williams, A. (2010). *The water footprint of English beef and lamb production, a report for EBLEX*. Cranfield, Bedfordshire: Cranfield University, Department of Natural Resources.

Chapagain, A. K., & Orr, S. (2009). An improved water footprint methodology linking global consumption to local water resources: A case of Spanish tomatoes. *Journal of environmental management*, *90*(2), 1219–1228. doi:10.1016/j.jenvman.2008.06.006.

Fader, M., et al. (2011). Internal and external green-blue agricultural water footprints of nations, and related water and land savings through trade. *Hydrology and Earth System Sciences Discussions, 8*, 483–527. doi:10.5194/hessd-8-483-2011.

Falkenmark, M. (1989). The massive water scarcity threatening Africa. Why isn't it being addressed. *Ambio*, *18*(2), 112–118.

FLAI-CGIL, edited by Osservatorio Placido Rizzotto (2012). *Agromafie e caporalato: primo rapporto*, CGIL.

Gerten, D., et al. (2011). Global water availability and requirements for future food production. *Journal of hydrometeorology*, *12*, 885–899. doi:http://dx.doi.org/10.1175/2011JHM1328.1.

Hoekstra, A. Y., & a cura di. (2003). Virtual water trade. In *Proceedings of the International Expert Meeting on Virtual Water Trade*. The Netherland: IHE Delft.

Hoekstra, A. Y., Chapagain, A. K., Aldaya, M. M., & Mekonnen, M. M. (2011). *The water footprint assessment manual: Setting the global standard.* US: Earthscan.

Hoekstra, A. Y., & Mekonnen M. M. (2012). The water footprint of humanity. *Proceedings of the national academy of sciences*, *109*(9).

Mekonnen, M. M., & Hoekstra, A. Y. (2011). National water footprint accounts: The green, blue and grey water footprint of production and consumption. *Value of water research report series no. 50*, UNESCO-IHE.

Molden, D., & IWMI. (2007). *Water for food. Water for life. A comprehensive assessment of water management in agriculture*. US: Earthscan.

Savenije, H., & van der Zaag, P. (2002). Water as an economic good and demand management. Paradigm with pitfalls. *Water international, 27*(1), 98–104.

Segrè, A., & Falasconi, L. (2012). *Libro blu dello spreco in Italia: l'acqua*, Edizioni Ambiente.

Sojamo, S., et al. (2012). Virtual water hegemony: The role of agribusiness in global water governance. *Water International, 37*(2), 169–182.

UNEP-DEWA. (2003). Groundwater and its susceptibility to degradation: a global assessment of the problem and options for management. *Early Warning and Assessment Report Series*, RS 03-3, Nairobi.

WWAP World Water Assessment Programme. (2012). *The United Nations world water development Report 4: Managing water under uncertainty and risk*, WWDR4 UNESCO, p. 3.

Water and Food Security: Food-water and Food Supply Value Chains

J.A. (Tony) Allan

Abstract The purpose of this chapter is to highlight the importance of food supply chains in understanding water security. Food supply chains are important because about 90 % of the water needed by an individual or a national economy is embedded in their food consumption. This water will be called food-water in this analysis. Food requires water to produce it. This water can be either green water—that is the water that is held in the soil profile after rainfall. Crops and vegetation can use this water for consumptive transpiration. Food-water can also be blue water, usually called freshwater. Such water can be diverted from rivers or pumped from groundwater. Globally, green water accounts for about 80 % of the water used for crop and livestock production. Over 20 % is blue water which is the water used consumptively in full and supplementary irrigation. The food supply chain is also important because farmers and other agents in this supply chain allocate and manage the vast volumes of water used consumptively. Farmers are helped by ag-industries which breed seeds and provide fertilizers, equipment and pesticides. All of these inputs plus science and many government subsidies have enabled farmers to increase their water productivity. Farmers manage about 90 % of the food-water resources in the food supply chain. The other 10 % of food-water is handled by corporations and other private sector entities that trade, transport, process and market food for consumers. The volumes of food-water in this non-farm part of the supply chain are therefore relatively small. (Note the analysis in this chapter does not address the water resources devoted to the production of fibre and energy. The author recognises the role of water in these economic activities but there is no space to address the nuances that these consumptive and non-consumptive demands place on the consumptive use of water.)

J.A. (Tony) Allan (✉)
King's College London, London University, London, UK
e-mail: ta1@soas.ac.uk

J.A. (Tony) Allan
SOAS, London University, London, UK

© Springer International Publishing Switzerland 2015
M. Antonelli and F. Greco (eds.), *The Water We Eat*,
Springer Water, DOI 10.1007/978-3-319-16393-2_2

1 Introduction—Hydro-system Fundamentals and Food Supply Chain Shaped Food-water Demands

> There are no water wars because food wars are not judged to be necessary.
>
> Society, politics and market players have conspired to put in place - globally and nationally - highly politicised global food regimes and food supply chains that have no reporting or accounting rules for water resources.

The purpose of the chapter is to highlight the importance of food supply chains in understanding water security. It will highlight both the politicised relationships, as well as the inescapable bond, between sustainable food security and sustainable water security. Sound food policies as well as sound water resource allocation and management will depend on the recognition of this connection.

The relationship between water and food is exceptional. No other supply chain needs or consumes a natural resource in the proportions that the world's food supply chains use water resources consumptively. The water used to produce food will in this analysis be called food-water.

Agents in the food supply chain have, since the beginning of farming about 13,000 years ago—usually unwittingly—been adapting and mainly enhancing the efficiency of the ways they mobilise the invisible rainfed green water in the root zone for food production. During all of this pre-industrial era, almost all the food-water consumed in crop transpiration was Nature's rainfed green water. Blue water —surface and groundwater—has been used in different modes of irrigated agriculture for the past five millennia. But very small volumes of blue water were consumed in irrigation until the beginning of industrialisation two centuries ago. The scale of the negative impacts on water resource ecosystems of humanity's industrial and post-industrial food supply chains has no precedent.

The analysis that follows will include the consideration of two major systems that bind together water resources and food production and consumption. First, there will be a very brief review of the significant characteristics of the hydro-systems that underpin the supply chains on which societies depend for their food security. It will be concluded that farmers are the major professionals that allocate and manage natural and engineered water resources. Secondly, there will be a very brief review of the history of global and other *food regimes* and *food supply chains*. The brief history of the world's food regimes shows that recent market volatility has exposed some dangerous features of the power asymmetries of the current global food regime. This regime emerged after the Cold War food regime of the 1950–1989 era. Thirdly, it will be shown that there exist many enduring sub-national food systems. These feed over 80 % of the world's populations (Hoekstra and Mekonnen 2012). Global food *trading* systems only ensure the food security of about 15 % of the global population (Hoekstra and Mekonnen 2012). This low proportion, however, belies its significance. The successful servicing of this international demand for traded food, driven by food consumption in water and food deficit economies, keeps the world at peace. It must be emphasised that it is normal to live in a food deficit economy. About 160 out of 210 economies in the world exist in conditions of inescapable food deficit of which they are usually innocent. There are no water wars because food wars are not judged to be necessary (Allan 2001).

There has been a clear shift in the nature of the global food system since the 1980s as a consequence of the expanded reach of global transnational corporations based mainly in the USA and Europe. Unfortunately, this third food regime adopted the most dangerous assumption of the first and second regimes, namely that that water is a free good. A feature of the past decade has been the emergence of some East Asian corporate traders with global ambitions. These developments may signal the beginning of a fourth global food regime (Keulertz 2012a).

The commercial and communication competence of these long established and some new corporations have led to an unprecedented concentration of market power (Williams 2012). The transnational corporations in these food supply value chains—often referred to as the brands and the non-brands—operate across the world. They can operate in short sub-national supply chains. Increasingly influential, however, are the long global food supply value chains which are very well integrated into the global food regime just discussed. These major players very well understand the operation of food supply chains and have well-developed information systems which uniquely privileges them in evaluating and handling environmental and market risks. They also have established and sometimes still own elements of the banking, hedging and insurance systems that underpin the operation of the supply chains. They stand out as the players who deploy huge influence.

Knowledge is certainly power. But the corporate knowledge is not framed on reporting and accounting rules that take the value and scarcity of water into account. Society, politics and these pivotal market players have conspired to put in place—globally and nationally—highly politicised food regimes and food supply chains that have no reporting or accounting rules for water resources. Food supply chains will always be highly politicised. They need to be re-politicised in a market economy that recognises the essential securitising roles of water and ecosystems.

2 The Important Underlying Hydro-fundamentals of Water Resources and Food Systems

> Society chooses to listen or not to listen to economists. It points accountants, not economists, at the operational problems of its markets when it wakes up to the dangers of exhausting a strategic input such as water or alienating a strategic input such as labour. Food-water is not yet a wake-up issue.

On the assumption that the readers of the chapter will mainly be those who produce, manufacture and consume food rather than those who work in the hydrosciences the terms used for water will be those that have proved to be accessible to the general reader since Falkenmark introduced them in 1986. Rather than those used by hydrologists, engineers and economists who plan and install water management infrastructures.

Society's food supply chains utilise about 90 % of the water used by society (Hoekstra and Mekonnen 2012). We shall call this water *food-water.* The other approximately 10 % is used by households and industry (Hoekstra and Mekonnen 2012). We shall call this municipal water *non-food-water*. Of the massive proportion of natural green and blue water embedded in the world's food supply chains, 90 % is used in producing food and fibre on farms by farmers. Farms are clearly where improved returns to water are delivered and water ecosystems are stewarded.

There is a third type of water—desalinated or manufactured water but this water is as yet unaffordable for crop production. Manufactured water is not considered in this analysis. Nor is the recycled water available after treatment of domestic and industrial use.

The two types of natural water used by farmers are green water or blue water. Globally, green water is the most important in terms of volume. About 80 % of the food in global farming is produced with *green water* (Hoekstra and Mekonnen 2012). *Green water* is the water that—during a cropping season enabled by rainfall events—stays long enough in the root zone to meet the evaporative needs of a harvestable crop. The metrics of the volumes of water available and of water transpired are both very difficult to capture (UNEP/GRID-Arendal 2009; UNEP/GRID-Arendal 2009). Global estimates that 80 % of green water is used to produce society's food production are usually based on estimates of crop transpiration modelled from remotely sensed multispectral data (Mulligan et al. 2011; Mulligan 2013). These metrics can also be augmented with estimates based on production data.

Green water has been very important throughout the 13 millennia of humanity's crop producing history. Despite the importance of green water in determining water and food security, it is only very recently that its role has been recognised and metrics developed. Such metrics are still not included in national and international water datasets such as those of FAO, although there are moves to remedy this situation (Margat et al. 2005).

The second type of water used in crop production is *blue water*. Blue water is first, the water present in surface flows and the water found in natural and man-made reservoirs. Secondly, it is the water in groundwater aquifers. While the massive volumes of green water have always been ignored blue water has generally been recognised as a vital resource once developed—because it is evident. Blue water is tangible. It can be pumped by engineers. Economists can conceptualise values for it. But both engineers and economists have failed to introduce concepts that move society and its food-water using food supply market chains to value or price blue water. Never mind green water. Society chooses to listen or not to listen to economists. It points accountants, not economists, at the operational problems of its markets when it wakes up to the dangers of exhausting a strategic input such as water or alienating a strategic input such as labour. Food-water is not yet a wake-up issue.

Despite blue water being evident and very widely revered as a holy cultural presence, it has rarely been valued as an economic input. Its value has not figured until its availability became scarce. Unfortunately by then it cannot be valued because that water resource use has become integral to unalterable livelihoods. These livelihoods are seen by society and by society's politics as unchangeable elements of an eternally politicised food economy.

The livelihoods that blue water has enabled are embedded in political economies created in a world that preceded the triple bottom-line assumptions of *people, profit* and *planet* (Elkington 1995) highlighted by activist scientists in the 1970s and 1980s. Adding the third environmental bottom line is an elemental political challenge as it is asking society to address the second failure of capitalism. In the two centuries of industrialisation since 1800—when society was being asked to address the first failure of capitalism—slavery and getting labour wrong, unprecedented demands for water have been imposed on blue water natural water ecosystems. These new water demands were mainly a consequence of the increase in the human population from about one billion to the current seven billions.

Fortunately, these past two centuries of industrialisation have also been associated with extraordinary increases in the productivity of food-water on farms (Allan 2011). The increased returns to both blue and green water have been a consequence of farmers combining increasing volumes of diverse inputs progressively more effectively. Farmers allocate and manage water resources more than any other profession. The main message of this chapter in relation to water resources and water resource security is that we depend on farmers individually and collectively. We depend on them now and will do so increasingly in the future. How they allocate and manage water will determine first, whether there will be sustainable volumes of clean water to produce our food needs. Secondly, if society enables farmers to have viable farm livelihoods, they will in turn be able to determine whether the water ecosystem services will be well enough stewarded to ensure their long-term viability. This long-term eco-viability—dependent on society recognising the role of farmers as water managers and stewards—will determine whether we can enjoy long-term socio-economic viability.

3 The Inescapable Political Economy Contexts—Global Food Systems and Food Supply Value Chains Contexts

> Human health and environmental and water ecosystem health have turned out to be very closely aligned.
>
> Society, politics and market players have conspired to put in place - globally and nationally - highly politicised global food regimes and food supply chains that have no reporting or accounting rules for water resources.
> 'It's true that you cannot talk yourself out of things you've behaved yourself into' (Polman of Unilever 2011).

This section will identify the second, political economy, underlying fundamental that must be understood by society if it is to manage water resources sustainably. This second sociopolitical condition is shaped by global food regimes and by numerous other food supply value chains developed by society. Both the long supply chains of the global food regime and those of the usually much shorter sub-national food supply value chains are dynamic. They sometimes experience very volatile phases as a number of factors impact and transform global food supply chains. Such a transformation has characterised the global nature and reach of the OECD long supply chains since 1989. It is also strongly impacting some of the BRICS economies and has long impacted many other economies. It is necessary to understand global food regimes and food supply chains more generally to grasp the significance of the collectively *water resource blind* forces that determine the way farmers manage water resources and whether they can steward them sustainably.

Food regime theory was developed by Friedmann (1978), Friedmann and McMichael (1989), McMichael (2009). It explains the underlying structures of the post-1950 structure of international agriculture. The 1939–1945 wars delivered an elemental shock to all the world's nations and particularly to the powerful warring parties. An unintended outcome was a strong post-war appetite for regulation. The 'regulation of the food regime both underpinned and reflected changing balances of power among states, as well as between organised national lobbies and classes—farmers, workers, peasants—and capital' (McMichael 2009). Food regime analysis notes that the globalisation of modern agriculture first became evident with the British-led outsourcing of agricultural activities to colonies in tropical zones and then to former-colonies more generally. The first regime was associated with the wealth transferring and humanitarian disaster of the imperial tropical sugar trade of the 1750–1850 period. It was succeeded between about 1870–1930 by an era of grain and livestock production and export from settler colonies to industrialised and rapidly urbanising European communities of all classes. This model functioned to feed the prospering middle classes around Europe with basic foodstuffs and with increasingly popular exotic commodities, such as tea and coffee. Interestingly most of the corporates now prominent in current global system had established themselves by 1870. The ABCD—global grain and livestock commodity traders (Murphy et al. 2012), and two of the biggest food brands—and Nestlé and Unilever were already established by the late nineteenth century. They have been joined by a much more numerous group of long-established US corporations including, Pepsi, Coca-Cola and Kellogg since the beginning of the twentieth century.

This first modern food regime lasted until the 1930s hiatus in global politics and especially in its international and national economic systems. After the 1939–1945 wars, leadership of the global food regime passed to the USA. The outcome was a system that prioritised arrangements that promoted a US Cold War agenda and the interests of US corporations. This US-led second modern food regime re-routed surplus food from North America to strategically important allies. Unfathomable incentives and subsidies were associated with the movement of cheap food commodities to reinforce loyalty against communism (Williams 2012). At the same time, very significant advances in agronomy, seed breeding, as well as in farm equipment, fertilizer and pesticide technologies led to the trebling of water productivity. Food commodity prices less and less reflected the costs inputs of natural resources and their misuse. These apparently economically beneficial advances were associated with the deluding mix of subsidies, such as under-priced inputs of energy and un-priced water. We are living with the impossible politics of reversing the beliefs, habits and mismanaging behaviour of the second global food regime. It put in place a toxic system made very dangerous indeed by the absence of water resources reporting and accounting systems.

The second 1950–1980 global food regime was, therefore, associated with a half century of misleading declines in global food commodity prices. The trend was everywhere welcome, especially by the poor in all types of economy. The politicians responsible for meeting the food needs of the poor were especially welcoming. The food commodity price spikes of the 1970s were, in retrospect, brief. Although one remembers as both a consumer and a scientist at the time being just as anxious then as in the middle of the current spasm of food price volatility. The outcome of the current 2008–2012 period of price volatility is hard to predict. Conditions are different from those of the 1970s. Populations and related demands have doubled and oil prices, which are the big brother of global commodity prices, are unlikely to fall to the levels of the 1980–2002 period.

The cheap food assumptions of the second food regime were doubly dangerous. First, they misled consumers and society about the real cost of food and water. Secondly, they encouraged consumers to waste food and water, as well as all the other costly inputs needed to produce food. Both rich and poor in both developing economies and in the OECD club were affected. The food choices of consumers in the rich economies especially ignored the impacts of their *wasteful* and *unhealthy* food choices on their own health and on the health of water ecosystems (SIWI and IWMI 2008). Human health and environmental and water ecosystem health have proved to be very closely aligned.

The US and European transnationals both adapted to, and were in a strong position to, promote the second regime. The process led among other things to a global agricultural division of labour with progressive commodification as part of a development strategy of Western states (McMichael 2009). At the same time, both the US and the European Union introduced a costly but in a number of ways a very efficient system of agricultural subsidies at least in protecting their markets against the perceived communist threat (Keulertz and Sojamo 2013).

The transnational corporates were also in a very favourable position when the third food regime emerged in the 1980s. This regime was initiated and enabled by what turned out to be a radical deregulating phase in the neo-liberal project driven by an influential state-market alliance of the political classes in North America and Europe. The transition was also strongly reinforced by the end of communism at the close of the 1980s. The critical dysfunctionality of this experiment was dramatically exposed in the crash and global economic hiatus of 2008. All global sectors and commodities—including the massive water consuming global food supply chains—were affected. These global food commodity markets are still experiencing price volatility that may not settle back as they did after the price spikes of the 1970s.

This powerful juggernaut—the integrated global food regime and the world's equally formidable food supply chains—is a political economy that remains water resource blind.

The cereal traders, the ABCD corporates have been very significant actors in this third post 1980s Western-led global food system and its supply chains (Sojamo 2010; Sojamo and Larson 2012; Murphy et al. 2012). Another important feature of this third global food regime has been the emergence of a new supermarket retailer and wholesaler nexus. The very rapid expansion of this supermarket system involved an unprecedented rationalisation of the food supply chains of the third food regime. For example, the already impressive advanced weather and market information systems developed by the corporate food commodity traders were massively reinforced by the rapid evolution of computerised global data-handling across the regime.

All the corporate food chain transnationals—brand and non-brand—benefitted from this *knowledge is power* revolution. Supermarkets, particularly, found themselves in an unprecedented position of market power that had the unintended consequence of exposing them to unwelcome reputational risk. In terms of water resource management, their new position meant that they were in an unfamiliar and vulnerable position. This vulnerability was highlighted by the appearance of the water footprint metrics first developed and published by Hoekstra and Hung (2002). The prominent and very reputation conscious brands—such as Tesco, Waitrose, Marks and Spenser in the UK and WalMart in the USA—were joined by their commercial partners in the global supply chain community—Unilever, Nestlé, Coca-Cola, Pepsi, Barilla (2012), SAB-Miller, in researching and publishing on the topic of embedded water and its environmental implications (WBCSD 2006, 2007; WWF 2012). Their business leaders moved unusually rapidly (UN Global Compact's CEO Water Mandate July 2007 (WBCSD 2006, 2007) to establish an intellectual lead as yet unmatched by the academy and the public sector. An example of the leadership being provided by the private sector is that of Jochen Zeitz who developed the *Combined financial and sustainability report* of *Puma* (2011).

A number of leading environmental NGOs kept pace, provided leadership and produced influential publications (WWF 2008; Waterwise 2007). The Water and Resources Action Programme (WRAP 2011) is another excellent example of how NGOs have worked with the private sector to identify the hot spots and impacts of natural resource misuse. WRAP is currently funding important research on weighted water footprints (URS 2013).

The private sector in food supply chains is always the major allocator and manager of natural resources. A complicating feature in the case of water resources is that private sector farmers in private sector food supply chains have been weak players in all three of the global food regimes. It is interesting that it is WRAP an NGO that has commissioned its *Environmental data and hotspot impact research* report (URS 2013) and not a department of state or a funded science research programme in the academy.

A leading position has been established by the transnational food corporates in the international discourse on responsible water resource use. The lead has been established with exceptional flair in a number of cases. Some CEO's of influential brands in the transnational global food regime and its food supply value chains have very effectively established themselves as lead advocates. They fluently coin persuasive ideas on stewarding water resources—'It's true that you cannot talk yourself out of things you've behaved yourself into' (Polman of Unilever 2011). 'The world is out of water already. It will not be able to grow as fast as it did in the past if we do not get a grip on the water side' (Nestlé's Brabeck-Lemanthe 2012). They and the supermarkets are constructing the priorities of the world's food supply chains. They have market power. They also have platforms in the publications of powerful ideas agencies such as McKinsey who assert that 'a complete rethink of resource management will be needed to keep pace with soaring demand as up to three billion new consumers join the world's middle classes over the next 20 years' (McKinsey 2012). At the World Economic Forum, they contribute to its agenda forums and to its declarations and publications (World Economic Forum 2011; UN Global Compact 2007). So far ahead are the leading supply chain brands in understanding the nature and challenges of water resources in the food supply chain that they can relax and wait for the academy and public sector policy-making to catch up.

With the propensity of markets to fail in environmentally challenging conditions, this asymmetric knowledge/power situation is not a promising scenario for the evolution of a secure and sustainable global food/water political economy. In a market system, *profit* will always be a tempting short-term default. The outcome for blind *people* and for the invisible *planet* is not promising in a world with unchallenged and very adept transnationals.

Meanwhile departments of state of the nations of the world and related public science and policy agencies reveal that they have not yet adopted the idea that our strategically important food-water is embedded in food supply chains. (DEFRA 2011; Australian Government 2012; Federal Ministry of Economic Cooperation and Development 2006). There is a little evidence that things may be changing at the national level. Recent studies commissioned by the US State Department note embedded water but they do not consider the political economy of food supply chains with respect to water security (Intelligence Community for the US State Department 2012). Numerous international agencies are also unaware of the role of food supply chains in water resource security (World Bank [on China] 2012; ERD [a European perspective] 2012; FAO 2013; FSDL 2012; UNESCO/UN-Water 2012; OECD 2012a, b) in their reporting on the state of the world's water resources. A number of UN/CGIAR agencies (FAO 2013; IWMI 2010; UNESCO/UN-Water 2012; World Bank 2012; OECD 2012b) have recognised that it is unavoidable that root-zone water from rainfall is an essential element of the global water budget. But they have neither devised metrics to get such water taken into account in the policy domain. Nor have they engaged with the unavoidable co-evolution that is the outcome of the encounter of such hydro-thinking and food-water allocation and management in farming. Nor have they engaged with the political economy of the global food regime and the supply chains which 'we have behaved ourselves into'.

4 Food-Water and the Markets of the Global Food Systems and its Food Supply Value Chains into Which We Have Behaved Ourselves

State and other food supply chain players have - understandably - preferred to align with society's cheap food fixation rather than with food-water security and water ecosystem stewardship.

There is no appetite for reporting and accounting rules.

We have identified five key issues. First, farmers manage and potentially steward society's food-water, that is 90 % of society's water. Secondly, society has not provided its farmers with the necessary resources to cope with environmental and market uncertainties to enable them to enjoy secure livelihoods never mind the new responsibility of stewarding the water ecosystems on which society itself depends. Thirdly, society—as consumers and voters—have preferred the availability of cheap food to addressing the second failure of capitalism—that is, failing to adopt measures that use natural resources such as water sustainably. The first failure was getting labour wrong two centuries ago. Fourthly, state and other food supply chain players have, understandably, preferred to align with society's cheap food fixation rather than with food-water security and water ecosystem stewardship. Fifthly, there is no appetite for reporting and accounting rules. This condition is partly because devising and installing them would be very complicated. But it is mainly because capturing the costs of food-water would impact the price of food, which is a neuralgic political issue for consumers, politicians and the food-water blind market food supply chains.

Food supply chain is an extraordinary market phenomenon. They use nearly all the water needed by society. In addition—Fig. 1 shows that—the agents in the whole supply chain—from farmers to the food commodity consumers—are all in the private sector. They produce, trade, manufacture, process, retail and consume food via private sector markets. This contrasts markedly with non-food-water services—these are the domestic and industrial water services—which are almost exclusively organised and delivered worldwide by the public sector. The UK is an exception. In Italy, privatising non-food-water services is a very highly politicised issue.

In the water intensive food supply chain, farm owning farmers and private corporations run everything from family farms to major transnational corporations that trade, manufacture and retail food. They all use accounting rules that are blind first, to the cost of food-water inputs and secondly, to the impact of farming, especially the associated misallocation and mismanagement of blue water in irrigation, on the ecosystem services of water.

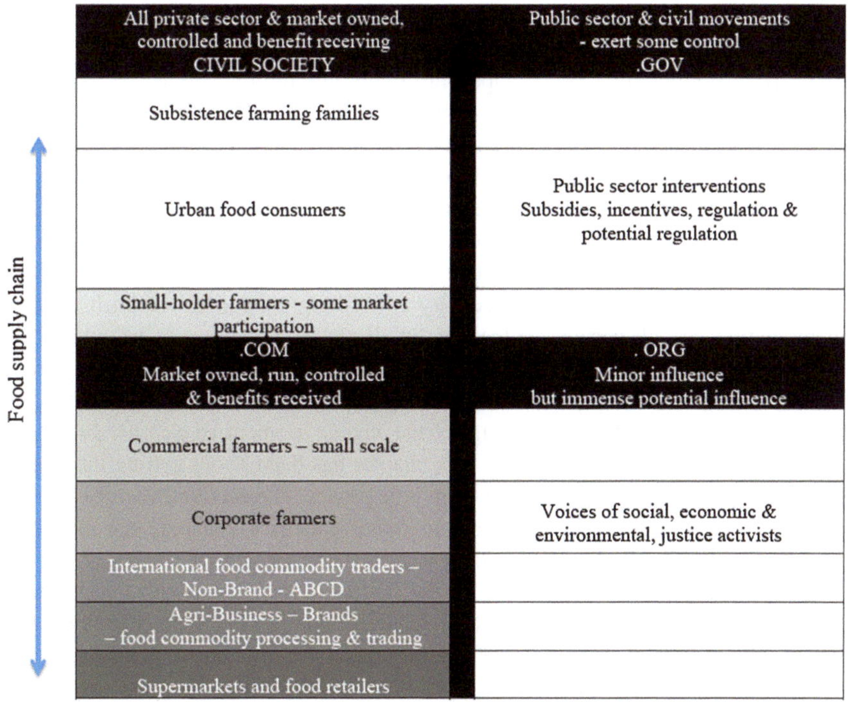

Fig. 1 An analytical structure of the agents and *social solidarities* (Douglas 1992) involved in the food supply chain, from food production by farmers to food consumption by individual consumers. The food-water in the private sector food supply chain—90 % of the water needed by an individual or economy—is allocated and managed by farmers and then supplied by agents in the private sector markets using water-value blind rules. *Source* Douglas (1992) and Allan (2013)

Figure 1 illustrates the phenomenon that this chapter has been emphasising that our very water intensive food supply chains are in the private sector. The agents who have long managed and still manage 90 % of natural green and blue water worldwide are on the left side of the diagram. They exist in civil society, first as food consumers and family farmers in the upper left solidarity and second as commercial and corporate farmers and as corporate entities in the rest of the food supply chain markets in the lower left solidarity. The players in the lower left solidarity have market power or potential market power. They can sometimes enjoy dominant roles. In contrast, farmers—the main water managers—may face impossible market conditions in dynamic circumstances that can include being driven out of business by their age, debt, pests, scarcity of affordable labour, drought, flood or negative market conditions including adverse international trade environments. All of these conditions preoccupy the farmer to the exclusion of the consideration of water resources. In the exceptional circumstances of advanced economies, crop insurance schemes have demonstrated that they can help farmers at the margin to survive. But these operational circumstances are exceptional.

States and those who make and deploy state power—in the top right solidarity—have over millennia put in place regulatory measures and incentives, tuned by very tough farm politics, that have shaped farm and agricultural policies as well environmental and trade policies. These evolving policies overlay very long-established rules of land ownership and tenure and labour rights and conditions. There is little, and usually no, appetite as yet to add to this list of regulatory measures by engaging in the toxic politics of capturing the value of water in food supply value chains.

Agricultural history is clear. Valuing an input in the food supply value chain is very challenging and involves conflictual politics by many generations over periods as long as a century. And even a century is not long enough to establish arrangements that satisfy all parties, all the time in all places. Processes of valuing labour and reforming labour rights initiated at the end of the eighteenth century took a century to put in place. It proved to be a long, ugly and sometimes violent, struggle to get markets to do the right thing by the *people* element of *people, profit and planet* triple bottom-line approach. This chapter has been about getting the market to do the right thing by the third element—the *planet*. The accounting rules in place are as yet very partial. They function to provide metrics that help the market to conjure the middle element—*profit*.

As yet the only solidarity—that in the bottom left quadrant—that has consistently promoted the value of water is the same one that rescued society from the first failure of capitalism. This solidarity is populated by civil movements, non-government organisations, trades unions and other social entities. They are inspired by environmental ethics and principles of sustainability. These organisations serve as the moral compasses of societies and of their state institutions and of their markets. A number of international NGOs have played a pivotal role in promoting the arguments presented in this chapter (WWF 2008, 2012; Waterwise 2007; Murphy et al. for Oxfam 2012; URS for WRAP 2013; Elkington 1995; World Economic Forum 2011). They have no state or market power. They can only deploy advocacy. But they have been very effective in using the language of risk. Highlighting reputational risk has been especially effective in impacting the strategic thinking of the leaders of major food manufacturing and supermarket brands in both national and global food supply value chains. The NGOs had to accept that they had no choice but to engage with the powerful private sector corporations and with private sector farmers. Unlike the academy and governments, they spotted what they needed to understand and whom to impact. They had to relate to those who both manage and misallocate water resources and potentially steward water ecosystems. They also get invited to join meetings in government departments on the rare occasions that these government agencies consider national water strategies.

Tinkering with, never mind confronting, any of the major alliances of the globalised state and market players is to engage with elemental global power relations. Big oil, big auto, big tobacco, big pharma, big armaments, big media and bi-food all have some rather evident and many more non-transparent relations with major OECD economies, with worrying asymmetric outcomes. The global food regime and food supply chains that have been highlighted in this chapter are just another such arrangement of major significance to those responsible for other aspects of national and global security.

The experience of big oil is relevant to those of us trying to predict what will happen next in the global food system. Big oil included seven major—USA, UK and UK/Dutch—transnationals during the 1973 and 1979 oil price spikes. These major oil companies had written global energy history from the beginning of the twentieth century up to 1973. This big-oil oligopoly decreed at the beginning of the twentieth century that the world price of oil should be US$2 per barrel; and it was so from 1900–1973 except in the exceptional circumstances of world wars. This global energy regime managed to provide the world with a further three decades of under-priced energy after 1979. Energy consumers became just as addicted to cheap energy as they had to cheap food. Both the global energy and the global food regimes had not priced in the proper cost of natural resource inputs nor of damaged ecosystems.

By the first decade of the twenty-first century, the seven major transnational oil and gas majors had merged into four and they no longer wrote the global agenda. They had lost control of the energy supply chains and no more determined the global energy regime. They also found themselves being forced to take the very technically and commercially challenging exploration concessions. They were increasingly underbid on the easy concessions by producer country companies and especially by less environmentally responsible BRICS operators. The former mighty majors were forced to pick up the high-risk and high-cost concessions. This outcome in turn involved coping with the contradiction that these contracts were usually high risk environmentally and proved to very high cost commercially.

No alternative dominant coordinated global energy power nexus such as the pre-1980 Seven Sisters big-oil regime has emerged. But it is certainly the case that the Western major oil companies have in less than three decades become a shadow of their former selves. At the same time, their essential ally the US Government is struggling to maintain its global hegemony in the oil and gas sectors. It has a motley suite of alliances with unstable Gulf oil economies and an uncertain domestic energy policy involving non-viable bio-fuels and a controversial oil-shale fracking industry.

Global food regimes have been around for much longer than the twentieth century big-oil oligopoly. Interestingly, the global food system is under-pinned by a group of staple-food suppliers that is much smaller than the current list of major oil and gas exporters. There are over twenty significant oil and gas producers accounting for most of the global oil and gas trade. In contrast, the strategically important global grain trade is dominated by only five major net-exporters of water intensive staples—the USA, Canada, Brazil, Argentina and Australia. They trade with another very small number of grain commodity traders based in the USA and France—the ABCD corporations, plus the Swiss-based Glencore, which is consolidating a number of smaller traders. They appear to operate a durable Western-based global state/market alliance. But, (Keulertz 2012c) has pointed out that there is a new acronym to consider. The long-established US-/French-aligned global ABCD phenomenon has been joined by an East Asian group of four global grain traders that aim to serve the needs and interests of mainly Asian net-grain importers—three are based in Singapore and one in Indonesia. They are the NOWS corporations—the Noble Group which grew by 25 % in 2011 (Keulertz 2012c), Olam, Wilmar and Sinar Mas. In 2011, they handled just over 20 % of the business transacted by that ABCD corporations but their trade is growing rapidly.

The global food regime could well be entering a new phase (Keulertz 2012b). The global alignments are likely to be subject to change. The ABCD traders have not yet adopted even the shared values vision of the brands such as Nestlé (Sojamo and Larson 2012). The NOWS corporations have no immediate incentive to adopt water value and water ecosystem aware systems. However, Olam is setting a principled natural resource awareness pace in its operations (Olam 2012).

5 Achieving Food-water Security—Revisiting the Politicization of Food and Water Security: Concluding Comments

> 'Farmers, Nature and don't waste food.' Japanese pre-meal vow/pledge.

> Legislators have been voted in by electorates who are as water resource are illiterate as themselves. They have to engage with farm lobbies most of whose members are indebted and some suicide prone and with transnationals with powers way beyond those that they themselves possess.

The analysis has focused on the unheralded and unvalued 90 % of water embedded in the world's food supply chains (Allan 2011; Reimer 2012). It has especially highlighted the role of farmers who manage 90 % of the food-water resources used to produce that food. It has been shown that food consumers in society and all those who operate its food supply chains have colluded in a water resource allocating and managing system, which has assumed throughout that water for food production should be free and can certainly be ignored. The other dangerous assumption that compounds the water resource catastrophes across the world is that food should be cheap. Governments want their poorest citizens to have access to cheap food. Private sector markets compete to provide it. No one—with the exception of NGO environmental and health activists and some journalists—wants to put accounting for water resource inputs and ecosystem impacts on the agenda.

The economic security of rural livelihoods associated with food production has predictably determined the nature of the political economies that provide underpriced food. Most farmers, the key water managers, are also poor. As a consequence, they cannot absorb the costs of being society's water ecosystem stewards. Farmers need help from society to do what society requires via a transformation of global food supply chains. Food consumers in society, politicians, those who operate food supply chains and the global food regime in which they are all embedded have not yet recognised the role that must be played by farmers in bringing about food security. That food security is totally wedded to water security is also unrecognised. The existing system is also a bag of contradictions. Beyond the farmer, the food supply chain players are fixated with their competition to provide cheap food. Cheap food can only be made available when water ecosystems are un-costed. Food consumers have yet to be persuaded of the imperatives in the Japanese prayer at the beginning of this section where both farmers and Nature are valued.

It has been shown that the analysts in government, the international agencies and the academy have been blinkered. They are still looking in the wrong place for who is capable of addressing the current water resource problems. They call on hydrologists and water resources science including specialists in water governance. Understanding water governance is very important. But it has to be a governance reform agenda that recognises private sector markets and their food supply value chains and global food systems.

The global energy supply chain has been referred to a number of times in the analysis. Like food commodities—at least in their global manifestation—oil and gas and its predecessor coal have always been explored for, sourced and marketed in private sector systems. Unlike water resources professionals, however, all those who attempted to evaluate and understand oil and gas assumed they would use data, information and analysis generated by the major transnational companies. Any international conference on energy at least until the 1990s would be mainly populated by private sector transnational professionals. The majority of contributors would be private sector specialists. Public policy-makers and the academy would always be playing catch-up and generally failing to develop national energy policies. It is apparently still the case as the UK department of state responsible for Energy and Climate Change (DECC) is currently recruiting expertise from the private sector to be able to engage in horizon scanning and scenario planning (Guardian 2012).

The contrast with the water sector is stark. Water resource professionals have been faced with the increasing hot-spot crises of water allocation and use for over three decades. But they have not in any serious way involved the major players in the private sector food supply chains. They do not relate to farmers over water despite farmers managing 90 % of society's water budget. International and other meetings on water resources and water resources security and the research that is reported still focus on water resources science supplemented by hydraulic engineering and some social science. For two decades, the economics of water and water resources governance have also figured on the agenda of such meetings. But such analysis and any related metrics have focused on underlying economic processes and on public policy and not on who is operating the big water using practices of the food supply chains. Economists do not operate food supply chains. These private sector market entities run on reporting and accounting rules installed by very highly politicised processes where economic principles are nowhere near the commanding heights of the debates. These politicised processes deliver outcomes that are the result of fiercely contended assumptions and interests that are very selective in what they prioritise (Paalberg 2010). Those engaged are elected legislators as well as a myriad of more or less effective lobbies. Legislators have been voted in by electorates who are as water resource illiterate as themselves. They have to engage with farm lobbies many of whose members are in debt and some are suicide prone and also with transnationals with powers way beyond those that they themselves possess.

The oldest lobby in the world is the farm lobby. For millennia, farmers have always gone straight to politicians. The reporting and accounting rules that emerge from these contentious politics (Paalberg 2010) shape private sector markets including those that use 90 % of the water resources in hydro-blind food supply value chains.

Now that we can recognise that these supply chains operate in a highly politicised landscape; it is time to face the political facts of life with respect to installing secure and sustainable allocation and use of water resources. The newly emerging global food regime and the world's food supply value chains require a transition to a new political economy that has reporting and accounting rules. These rules would help ensure that farmers have viable livelihoods that in turn enable them to steward water ecosystems effectively. Water resource professionals and scientists both in the academy and government need urgently to recognise that water resource crises will be solved by farmers who have environmentally and commercially responsible contractual relations with their supply chain customers. Farmers and accountants will solve water resource allocation problems. But only if society, its food consumers as well as its global food supply chains, its legislators and its scientists can get their ideas properly aligned. This nexus needs to be revisited and differently politicised.

References

Allan, J. A. (2001). *Virtual water: Hydropolitics and the global economy*. London: I B Tauris.

Allan, J. A. (2011). *Virtual water: Tackling the threat to the planet's most precious resource*. London: I B Tauris.

Allan, J. A. (2013). Food-water security: Beyond water resources and the water sector. In B. Lankford, K. Bakker, M. Zeitoun, & D. Conway (Eds.), *Water security: principles, perspectives, practices*. London: Earthscan.

Australian Government. (2012). Australia's water, Canberra: Department of sustainability, environment, water, population and communities.

Brabeck-Lemanthe, P. (2012). http://www.huffingtonpost.com/2012/07/12/peter-brabeck-nestle-water-crisis_n_1667816.html

DEFRA. (2011). *Future water: The government's water strategy for England and Wales*. London: DEFRA.

Douglas, M. (1992). *Risk and blame: Essays in cultural theory*. London, New York: Routledge.

Elkington, J. (1995). People, profit and planet. *SustainAbility*. http://www.sustainability.com/

ERD. (2012). *Confronting scarcity: Managing water, energy and land*. Brussels: European Development Report. http://www.erd-report.eu/erd/report_2011/report.htm

Falkenmark, M. (1986). Fresh water—Time for a modified approach. *Ambio, 15*(4), 192–200.

FAO. (2012). Rome. Available online: http://www.fao.org/nr/tenure/voluntary-guidelines/en/

FAO. (2013). *Water resources strategies, in preparation*. Rome: FAO.

FSDL. (2012). *Outcome report of the conference on food security in dry lands.* Doha: QNSFP.
Federal Ministry of Economic Cooperation and Development. (2006). *Water sector strategy.* Berlin: Federal Ministry of Economic Cooperation and Development.
Friedmann, H. (1978). World market, state and family farm: Social bases of household production in an era of wage-labour. *Comparative Studies in Society and History, 20*(4), 545–586.
Friedmann, H., & McMichael, P. (1989). Agriculture and the state system: The rise and fall of national agricultures, 1870 to the present. *Sociologia Ruralis, 29*(2), 93–117.
Guardian. (2012). Energy company staff work at climate change ministry, The Guardian. Almost two dozen employees from companies including the energy giants British Gas and nPower are working at the Department of Energy and, in most cases, are being paid by the government to do so, documents released under freedom of information rules reveal. Oil companies such as Shell and Conoco-Phillips also have staff inside the department, and civil servants have travelled in the opposite direction to work for the companies. The DECC spokeswoman said: We ensure any secondee is bound by relevant professional codes of practice. Contractual measures make sure that any secondee is not placed in a position where they could be a conflict of interest. December 30, 2012.
Hoekstra, A. Y., & Hung, P. Q. (2002). The quantification of virtual water flows between nations with respect to crop trade, value of water research project. Report number 11.
Hoekstra, A. Y., & Mekonnen, M. M. (2012). *PNAS, 109*(9), 3232–3237.
Intelligence Community Assessment (ICA). (2012). *Global water security, a report requested by the US State Department.* Washington DC: ICA.
IWMI. (2010). *Managing water for rainfed agriculture, IWMI water issue brief.* Colombo: IWMI.
Keulertz, K. (2012a). Land grabs and the green economy. In J. A. Allan, M. Keulertz, S. Sojamo, & J. Warner (Eds.). *Handbook of land and water grabs in Africa: foreign direct investment and food and water security* (pp 243–336). London: Routledge.
Keulertz, M. (2012b). *The Middle Eastern food security question and the global food regime. Food security in dry lands*, Doha: Qatar Food Security Program.
Keulertz, M. (2012c) *The Sudanese breadbasket: Land and water grabs by Middle Eastern economies.* Drivers of the Arab rush for land, Presentation at the Oxford University African Studies Centre, December 2012.
Keulertz, M., & Sojamo, S. (2013). Inverse globalisation? The global agricultural trade system and Asian investments in African land and water resources. In *Handbook of land and water grabs: foreign direct investment and food and water security* pp 324–334. Abingdon: Routledge.
Margat, J., Franken, K., & Fuarés, J.-M. (2005). Key water resources statistics in AQUASTAT: FAO's global information system on water and agriculture. Intersecretariat working group on environment statistics (IWG-Env), international work session on water statistics, Vienna, June 20–22, 2005.
McKinsey. (2012). *McKinsey on sustainability & resource productivity.* http://www.mckinsey.com/client_service/sustainability/latest_thinking/mckinsey_on_sustainability.
McMichael, P. (2009). A food regime genealogy. *Journal of Peasant Studies, 36*(1), 139–169.
Mulligan, M. (2013). The water resource implications for and of FDI projects in Africa: A biophysical analysis of opportunity and risk. In *Handbook of land and water grabs: Foreign direct investment and food and water security.* Abingdo: Routledge.
Mulligan, M., Saenz Cruz, L.L., Pena-Arancibia, J., Pandey, B., Mahé, G., & Fisher, M. (2011). Water availability and use across the challenge program on water and food (CPWF) basins. *Water International, 36*(1), 17–41. http://dx.doi.org/10.1080/02508060.2011.543801
Murphy, S., Burch, D., & Clapp, J. (2012). *Cereal secrets: The world's largest grain traders and global agriculture.* Oxford: Oxfam Research Reports.
OECD. (2012a). *OECD environmental outlook to 2050: The consequences of inaction.* Paris: OECD Publishing.
OECD. (2012b). Water outlook to 2050: The OECD calls for early and strategic action. In *Global water forum,* Marseilles 2011. http://www.globalwaterforum.org/2012/05/21/water-outlook-to-2050-the-oecd-calls-for-early-and-strategic-action/.
Olam. (2012). *Olam corporate responsibility report.* Singapore: Olam.

Paalberg, R. (2010). *Food politics*. Oxford: Oxford University Press.

Polman, P. (2011). *McKinsey conversations with global leaders: Paul Polman of Unilever*. New York: McKinsey Quarterly.

Puma. (2011). *Combined financial and sustainability report 2011*. Munich: Puma.

Reimer, J. J. (2012). On the economics of virtual water trade. *Ecological Economics, 75*, 135–139.

SIWI & IWMI. (2008). *Saving water: From field to fork—Curbing losses and wastage in the food chain, SIWI policy brief*. Stockholm: SIWI.

Sojamo, S. (2010). *'Merchants' of virtual water: The 'ABCD' of agribusiness TNCs and global water security*. Unpublished M.Sc. dissertation, Department of geography, King's College London, London.

Sojamo, S., & Larson, E. A. (2012). Investigating food and agribusiness corporations as global water security, management and governance agents: The case of Nestlé. *Bunge and Cargill, Water Alternatives, 5*(3), 619–635.

UNEP/GRID-Arendal. (2009). 'Water scarcity index', UNEP/GRID-Arendal maps and graphics library (online). Available: http://maps.grida.no/go/graphic/waterscarcity-index. Accessed 6 Jan 2011.

UNESCO/UN-Water. (2012). *Managing water under uncertainty and risk*. Paris: UNESCO.

UN Global Compact. (2007). *CEO's water mandate, is a unique public-private initiative designed to assist companies in the development, implementation and disclosure of water sustainability policies and practices*. New York: United Nations (powerg@un.org).

URS. (2013). *Environmental data and hotspot impact research—Water metric feedback report*. A report prepared for WRAP (waste and resources action programme (UK)). Manchester: URS.

Waterwise. (2007). *Hidden waters, a waterwise briefing by Joanne Zygmunt*. London: Waterwise.

WBCSD. (2006). *Business in the world of water: WBCSD water scenarios to 2025*. Geneva: WBCSD.

WBCSD. (2007). *Global water tool*. CERES and Pacific Institute: CDP Water Initiative.

World Business Council for Sustainable Development. (2012). Global water footprint tool. Online: www.ceowatermandate.org

Williams, J. (2012). *Competition and efficiency in international food supply chains: Improving food security*. Abingdon: Routledge.

World Bank. (2012). Addressing water scarcity in China. China's leadership is well aware of the worsening water shortage and is determined to transform China into a water-saving society through policy and institutional reforms. World Bank 2012.

World Economic Forum. (2011). *Water security: The water-energy-food-climate nexus*. Washington D.C: Island Press.

WRAP. (2011). *New estimates for household food waste in the UK*. London: WRAP.

WWF. (2008). *UK water footprint: The impact of the UK's food and fibre consumption on global water resources*. Leatherhead: WWF.

WWF. (2012). *Assessing water risk: A practical approach for financial institutions*. Leatherhead: WWF.

The Water Footprint: The Relation Between Human Consumption and Water Use

Arjen Y. Hoekstra

Abstract It is increasingly recognised that freshwater scarcity and pollution are to be understood in a global context. Local water depletion and pollution are often closely tied to the structure of the global economy. With increasing trade between nations and continents, water is more frequently used to produce export goods. International trade in commodities implies long-distance transfers of water in virtual form, where virtual water is understood as the volume of water that has been used to produce a commodity and that is thus virtually embedded in it. Knowledge about the virtual water flows entering and leaving a country can cast a completely new light on the actual water scarcity of a country. At the same time, it becomes increasingly relevant to consider the linkages between consumer goods and impacts on freshwater systems. This can improve our understanding of the processes that drive changes imposed on freshwater systems and help to develop policies of wise water governance. The water footprint is an innovative concept to analyse water consumption and pollution along supply chains, assess the sustainability of water use and explore where and how water use can best be reduced. This chapter shows how the water footprint concept can be used to understand the international dimension of water and to assess water use behind daily consumer goods. This chapter argues for greater product transparency, water footprint ceilings per river basin and water footprint benchmarks for water-intensive commodities.

1 Introduction

The desirability of reducing our carbon footprint is generally recognised, but the related and equally urgent need to reduce our water footprint is often overlooked. Recent research has shown that about 4 % of the water footprint of humanity relates to water use at home (Hoekstra and Mekonnen 2012). This means that if people

A.Y. Hoekstra (✉)
University of Twente, Enschede, The Netherlands
e-mail: a.y.hoekstra@utwente.nl

© Springer International Publishing Switzerland 2015

M. Antonelli and F. Greco (eds.), *The Water We Eat*,
Springer Water, DOI 10.1007/978-3-319-16393-2_3

consider reducing their water footprint, they better critically look at their indirect water footprint than at their water use in the kitchen, bathroom and garden. Wasting water never makes sense, so saving water at home when possible is certainly advisable, but when we would limit our actions to water reductions at home, many of the most severe water problems in the world would hardly be lessened. The water in the Murray-Darling basin in Australia is so scarce mostly because of water use in irrigated agriculture (Pittock and Connell 2010). The Ogallala Aquifer in the American Midwest is gradually being depleted because of water abstractions for the irrigation of crops such as maize and wheat (McGuire 2007). In Italy, groundwater reservoirs in the south are overexploited, among others for growing durum wheat for making pasta (Aldaya and Hoekstra 2010).

In this chapter, I will first introduce the water footprint concept, an indicator increasingly used worldwide to assess the implications of consumption and trade on water resource use and pollution. Next, I will report on the water footprint of Italian consumption. Subsequently, I will zoom in on an important component in the water footprint of humanity: the hidden water resource use behind meat and dairy. I will compare the water footprint of animal products with the water footprint of crops and the water footprint of a meat-based diet with the water footprint of a vegetarian diet. I will then show that understanding the relation between food consumption and the use of freshwater resources is no longer just a local issue. Water has become a global resource, whereby—due to international trade—food consumption in one place often affects the water demand in another place. Finally, an argument is made for product transparency, which would allow us to better link individual products to associated water impacts, which in turn can drive efforts to reduce those impacts.

2 The Water Footprint Concept

The water footprint concept is an indicator of water use in relation to consumer goods (Hoekstra et al. 2011). The concept is an analogue to the ecological and the carbon footprint, but indicates water use instead of land or fossil energy use. The water footprint of a product is the volume of freshwater used to produce the product, measured over the various steps of the production chain. Water use is measured in terms of water volumes consumed or polluted. Water consumption refers to water evaporated or incorporated into a product. The water footprint is a geographically explicit indicator that shows volumes of water use and pollution, but also the locations. A water footprint generally breaks down into three components: the blue, green and grey water footprint. The blue water footprint is the volume of freshwater that is evaporated from the global blue water resources (surface and ground water). The green water footprint is the volume of water evaporated from the global green water resources (rainwater stored in the soil). The grey water footprint is the volume of polluted water, which is quantified as the volume of water that is required to dilute pollutants to such an extent that the quality of the ambient water remains above agreed water quality standards (Hoekstra and Chapagain

2008). In order to make sure that scientifically robust methods are applied and to ensure that a fair comparison can be made between different water footprint studies, the Water Footprint Network with its partners has developed the Global Water Footprint Standard, which was launched in February 2011 (Hoekstra et al. 2011). The figures presented in this paper are based on this standard. The Global Water Footprint Standard covers a comprehensive set of definitions and methods for water footprint accounting. It shows how green, blue and grey water footprints are calculated for individual processes and products, as well as for consumers, nations and businesses. It also includes methods for water footprint sustainability assessment and a library of water footprint response options.

3 The Water Footprint of Italian Consumption

On average, the water footprint of Italian consumption is 6300 L per day per person, which is 1.65 times larger than the global average. Only 4 % of the water footprint of Italian consumption is related to water use at home, which is in line with the global picture. About 96 % of the water footprint of consumption is thus 'invisible' for the consumer: it relates to the water consumption and pollution behind the products that consumers buy in the supermarket or elsewhere. About 89 % of the Italian water footprint relates to consumption of agricultural products and 7 % to industrial products. Nearly, half of the water footprint of Italian consumption is related to the consumption of animal products (Fig. 1).

About 60 % of the water footprint of Italian consumption lies outside the country. Figures 2, 3, 4 and 5 map the green, blue, grey and total water footprint of Italian consumption in the world. The largest fractions of the external water footprint of Italian consumption lie in France (harbouring about 9 % of Italy's external water footprint, mostly for production of animal products and wheat), Brazil (7 %, mostly animal products, soya bean, coffee), Germany (6 %, mostly animal products), Tunisia (6 %, mostly olives and cotton) and Spain (6 %, mostly olives). Next in row are the following countries: the USA (wheat, soya bean, animal products), Argentina (soya bean), India (cotton and coffee), the Russian Federation (wheat, animal products, sunflower seed, industrial products), the Netherlands (animal products), Romania (cotton, animal products, industrial products) and China (cotton and industrial products).

4 The Water Footprint of Animal Products

Much of the grains cultivated in the world are not for human consumption but for animals. In the period 2001–2007, on average 37 % of the cereals produced in the world were used for animal feed (FAO 2011). Surprisingly, however, there is little attention among scientists or policy makers to the relation between water use and

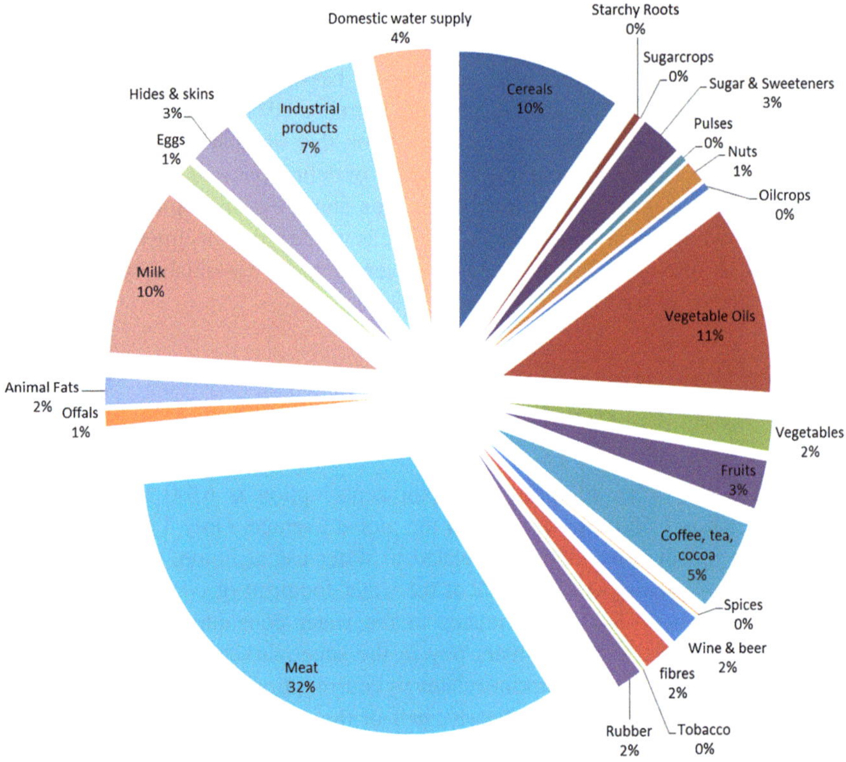

Fig. 1 The composition of the water footprint of the average Italian consumer. Period 1996–2005. *Data source* Mekonnen and Hoekstra (2011)

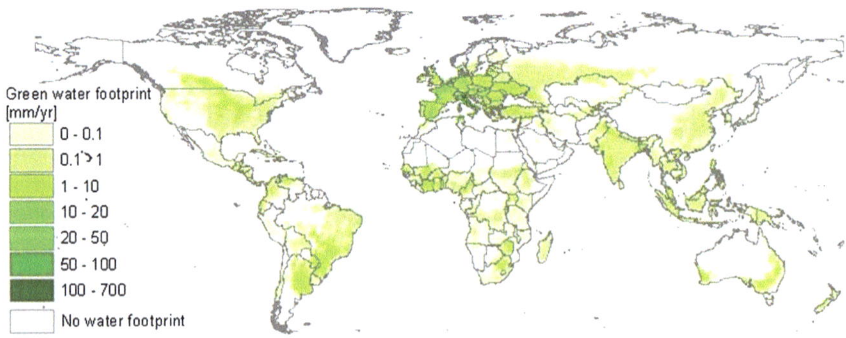

Fig. 2 The global green water footprint of Italian consumption. Period 1996–2005. *Source* Mekonnen and Hoekstra (2011)

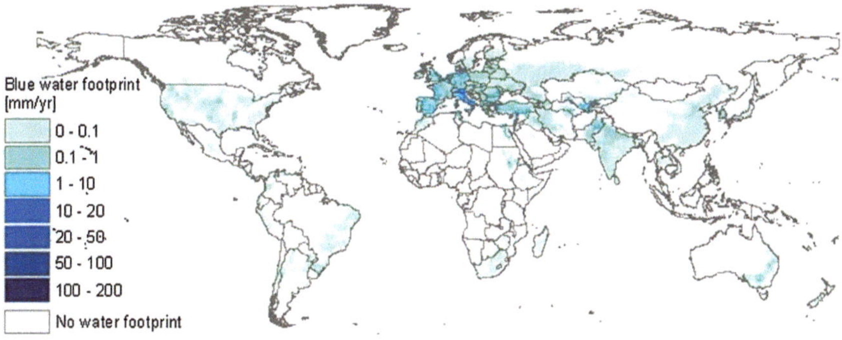

Fig. 3 The global blue water footprint of Italian consumption. Period 1996–2005. *Source* Mekonnen and Hoekstra (2011)

Fig. 4 The global grey water footprint of Italian consumption. Period 1996–2005. *Source* Mekonnen and Hoekstra (2011)

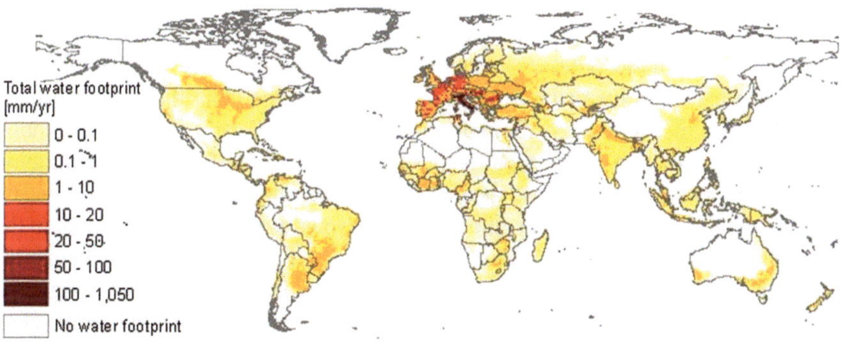

Fig. 5 The aggregated (green + blue + grey) water footprint of Italian consumption. Period 1996–2005. *Source* Mekonnen and Hoekstra (2011)

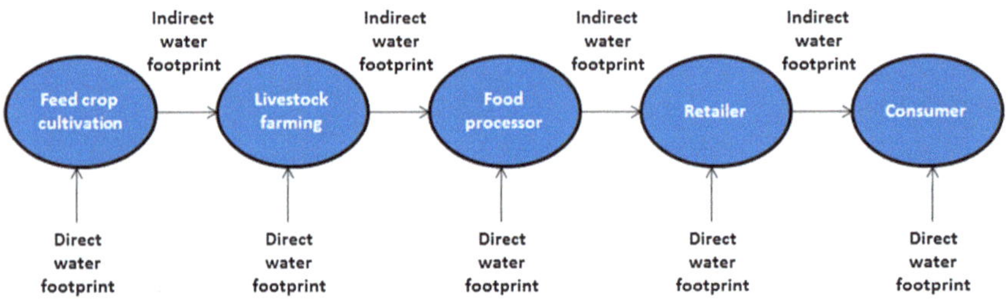

Fig. 6 The direct and indirect water footprint in each stage of the supply chain of an animal product. *Source* Hoekstra (2013)

the consumption of meat and dairy. It becomes increasingly relevant to study the implications of farm animals on water resource use, not only because the global meat production has almost doubled in the period 1980–2004 (FAO 2005), but also because of the projected doubling of meat production in the period 2000–2050 (Steinfeld et al. 2006).

The supply chain of an animal product starts with feed crop cultivation and ends with the consumer (Fig. 6). In each step of the chain, there is a direct water footprint, which refers to the water consumption or pollution in that step, but also an indirect water footprint, which refers to the water consumption or pollution in the previous steps. By far, the biggest contribution to the total water footprint of all final animal products comes from the first step: growing the feed. This step is the farthest removed from the consumer, which explains why consumers generally have little notion about the fact that animal products require a lot of land and water (Naylor et al. 2005). Besides, the feed will often be grown in areas completely different from where the consumption of the final product takes place.

In order to better understand the water footprint of an animal product, we better start with the water footprint of feed crops. The combined green–blue water footprint of a crop (in m^3/ton) when harvested from the field is equal to the total evapotranspiration from the crop field during the growing period (m^3/ha) divided by the crop yield (ton/ha). The crop water use depends on the crop water requirement on the one hand and the actual soil water available on the other hand. Soil water is replenished either naturally through rainwater or artificially through irrigation water. The crop water requirement is the total water needed for evapotranspiration under ideal growth conditions, measured from planting to harvest. It obviously depends on the type of crop and climate. Actual water use by the crop is equal to the crop water requirement if rainwater is sufficient or if shortages are supplemented through irrigation. In the case of rainwater deficiency and the absence of irrigation, actual crop water use is equal to effective rainfall. The green water footprint refers to the part of the crop water requirement met through rainfall; the blue water footprint is the part of the crop water requirement met through irrigation. The grey water footprint is calculated as the load of a pollutant (fertiliser, pesticide) that leaches from the field to the groundwater or runs off to the surface water divided by the difference between the maximum allowable and natural concentration for the pollutant in the water body.

The water footprint of an animal at the end of its lifetime can be calculated based on the water footprint of all feed consumed during its lifetime and the volumes of water consumed for drinking and for example cleaning the stables. One will have to know the age of the animal when slaughtered and the diet of the animal during its various stages of life. The water footprint of the animal as a whole is allocated to the different products that are derived from the animal. This allocation is done on the basis of the relative values of the various animal products, as can be calculated from the market prices of the different products. The allocation is done such that there is no double counting and that the largest shares of the total water input are assigned to the high-value products and smaller shares to the low-value products.

About 98 % of the water footprint of animal products relates to water use for feed (Mekonnen and Hoekstra 2012). A recent study by Gerbens-Leenes et al. (2011) shows that there are two major determining factors for the water footprint of animal products. The first factor is the feed conversion efficiency, which measures the amount of feed to produce a given amount of meat, eggs or milk. As animals are generally able to move more and take longer to reach slaughter weight in grazing systems, they consume a greater proportion of food to convert to meat. Due to this, the feed conversion efficiency improves from grazing systems through mixed systems to industrial systems and leads to a smaller water footprint in industrial systems. The second factor is the composition of the feed eaten by the animals in each system and works precisely in the other direction, in favour of grazing systems. When the amount of feed concentrates increases, the water footprint will increase as well, because feed concentrates have a relatively large water footprint, while roughages (grass, crop residues and fodder crops) have a relatively small water footprint. The increasing fraction of animal feed concentrates and decreasing fraction of roughages from grazing through mixed to industrial systems (Hendy et al. 1995) result in a smaller water footprint in grazing and mixed systems compared to industrial systems. In general, the water footprint of concentrates is five times larger than the water footprint of roughages. While the total mixture of roughages has a water footprint of around 200 m^3/tonne (global average), this is about 1000 m^3/tonne for the package of ingredients contained in concentrates. As roughages are mainly rain fed and crops for concentrates are often irrigated and fertilised, the blue and grey water footprint of concentrates are even 43 and 61 times that of roughages, respectively.

If we take beef as an example, it is clear from the above that the water footprint will strongly vary depending on the production region, feed composition and origin of the feed ingredients. The water footprint of beef from an industrial system may partly refer to irrigation water (blue water) to grow feed in an area remote from where the cow is raised. This can be an area where water is abundantly available, but it may also be an area where water is scarce and where minimum environmental flow requirements are not met due to overdraft. The water footprint of beef from a grazing system will mostly refer to green water used in nearby pastures. If the pastures used are either dry- or wetlands that cannot be used for crop cultivation, the green water flow turned into meat could not have been used to produce food crops instead. If, however, the pastures can be substituted by cropland, the green water allocated to meat production is no longer available for food crop production. This explains why the water footprint is to be seen as a multidimensional indicator. One should not only look at the total water footprint as a volumetric value, but also consider the green, blue and grey components separately and look at where each of the water footprint components is located. The social and ecological impacts of water use at a certain location depend on the scarcity and alternative uses of water at that location.

5 The Water Footprint of Animal Products Versus Crop Products

Mekonnen and Hoekstra (2012) have shown that the water footprint of any animal product is larger than the water footprint of a wisely chosen crop product with equivalent nutritional value. Ercin et al. (2011) illustrate this by comparing the water footprint of two soya products with two equivalent animal products. They calculate that 1 L of soya milk produced in Belgium has a water footprint of about 300 L, whereas the water footprint of 1 L of cow's milk is more than three times bigger. The water footprint of a 150-g soya burger produced in the Netherlands appears to be about 160 L, while the water footprint of an average 150-g beef burger is nearly fifteen times bigger. Table 1 shows the global average water footprint of a number of crop and animal products. The numbers show that the average water footprint per calorie for beef is twenty times larger than for cereals and starchy roots. The water footprint per gram of protein for milk, eggs and chicken meat is about 1.5 times larger than for pulses. For beef, the water footprint per gram of protein is 6 times larger than for pulses. Butter has a relatively small water footprint per gram of fat, even lower than for oil crops, but all other animal products have larger water footprints per gram of fat when compared to oil crops.

The global water footprint of animal production amounts to 2422 billion m^3 per year (87 % green, 6 % blue, 7 % grey). One-third of this total is related to beef cattle, another 19 % to dairy cattle (Mekonnen and Hoekstra 2012). The largest fraction (98 %) of the water footprint of animal products refers to the water footprint of the feed for the animals. Drinking water for the animals, service water and feed mixing water accounts for 1.1, 0.8 and 0.03 %, respectively.

6 The Water Footprint of a Meat Versus Vegetarian Diet

Dietary habits greatly influence the overall water footprint of people. In industrialised countries, the average calorie consumption is about 3400 kcal per day (FAO 2011), roughly 30 % of that comes from animal products. When we assume

Table 1 The global average water footprint of crop and animal products

Food item	Water footprint per unit of weight (L/kg)				Nutritional content			Water footprint per unit of nutritional value		
	Green	Blue	Grey	Total	Calorie (kcal/kg)	Protein (g/kg)	Fat (g/kg)	Calorie (L/kcal)	Protein (L/g protein)	Fat (L/g fat)
Sugar crops	130	52	15	197	285	0.0	0.0	0.69	0.0	0.0
Vegetables	194	43	85	322	240	12	2.1	1.34	26	154
Starchy roots	327	16	43	387	827	13	1.7	0.47	31	226
Fruits	726	147	89	962	460	5.3	2.8	2.09	180	348
Cereals	1232	228	184	1644	3208	80	15	0.51	21	112
Oil crops	2023	220	121	2364	2908	146	209	0.81	16	11
Pulses	3180	141	734	4055	3412	215	23	1.19	19	180
Nuts	7016	1367	680	9063	2500	65	193	3.63	139	47
Milk	863	86	72	1020	560	33	31	1.82	31	33
Eggs	2592	244	429	3265	1425	111	100	2.29	29	33
Chicken meat	3545	313	467	4325	1440	127	100	3.00	34	43
Butter	4695	465	393	5553	7692	0.0	872	0.72	0.0	6.4
Pig meat	4907	459	622	5988	2786	105	259	2.15	57	23
Sheep/goat meat	8253	457	53	8763	2059	139	163	4.25	63	54
Bovine meat	14414	550	451	15415	1513	138	101	10.19	112	153

Source Mekonnen and Hoekstra (2012)

Table 2 The water footprint of two different diets in industrialised countries

	Meat diet			Vegetarian diet		
	kcal/day	litre/kcal	litre/day	kcal/day	litre/kcal	litre/day
Animal origin	950	2.5	2375	300	2.5	750
Vegetable origin	2450	0.5	1225	3100	0.5	1550
Total	3400		3600	3400		2300

Source Hoekstra (2012)

that the average daily portion of animal products is a reasonable mix of beef, pork, poultry, fish, eggs and dairy products, we can estimate that 1 kcal of animal product requires roughly 2.5 L of water on average. Products from vegetable origin, on the other hand, require roughly 0.5 L of water per kcal, this time assuming a reasonable mix of cereals, pulses, roots, fruit and vegetables. Under these circumstances, producing the food for one day takes 3600 L of water (Table 2). For the vegetarian diet, we assume that a smaller fraction is of animal origin (not zero, because of the dairy products still consumed), but keep all other factors equal. This reduces the food-related water footprint to 2300 L/day, which means a reduction of 36 %. Keeping in mind that for the 'meat eater', we had taken the average diet of a whole population and that meat consumption varies within a population; larger water savings can be achieved by individuals that eat more meat than the average person.

From the above figures, it is obvious that consumers can reduce their water footprint through reducing the volume of their meat consumption. Alternatively, however, or in addition, consumers can reduce their water footprint by being more selective in the choice of which piece of meat they pick. Chickens are less water intensive than cows and beef from one production system cannot be compared in terms of associated water impacts to beef from another production system. Grazing livestock depends on local rain, while factory-farm livestock often relates to blue water consumption and pollution elsewhere.

7 The Local and Global Dimensions of Water Governance

Problems of water scarcity and pollution always become manifest locally and during specific parts of the year. However, research on the relation between consumption, trade and water resource use during the past decade has made clear that protection of freshwater resources can no longer be regarded as just an issue for individual countries or river basins. Although in many countries most of the food still originates from the country itself, substantial volumes of food, feed and animal products are internationally traded. As a result, all countries import and export water in virtual form, i.e. in the form of agricultural commodities (Hoekstra and Chapagain 2008; Allan 2011). Total international virtual water flows related to global trade in animal products add up to 272 billion m^3/year, a volume equivalent to about half the annual Mississippi run-off (Mekonnen and Hoekstra 2011).

Not only livestock and livestock products are internationally traded, also feed crops are traded (Galloway et al. 2007). In trade statistics, however, it is difficult to distinguish between food and feed crops, because they are mostly the same crops, only the application is different. Worldwide, trade in crops and crop products results in international virtual water flows that add up to 1766 billion m^3/year (Mekonnen and Hoekstra 2011).

Until today, water is still mostly regarded as a local or regional resource, to be managed preferably at catchment or river basin level. However, this approach obscures the fact that many water problems are related to remote consumption elsewhere. Water problems are an intrinsic part of the world's economic structure in which water scarcity is not translated into costs to either producers or consumers; as a result, there are many places where water resources are depleted or polluted, with producers and consumers along the supply chain benefiting at the cost of local communities and ecosystems. It is unlikely that consumption and trade are sustainable if they are accompanied by water depletion or pollution somewhere along the supply chain. Typical products that can often be associated with remote water depletion and pollution are cotton and sugar products. For animal products, it is much more difficult to tell whether they relate to such problems, because animals are often fed with a variety of feed ingredients and feed supply chains are difficult to trace. So unless we have milk, cheese, eggs or meat from an animal that was raised locally and that grazed locally or was otherwise fed with locally grown stuff, it is hard to say something about which claim such product has put on the world's scarce freshwater resources. The increasing complexity of our food system in general and the animal product system in particular hides the existing links between the food we buy and the resource use and associated impacts that underlie it.

8 Product Transparency

In order to know what we eat, we will need a form of product transparency that is currently completely lacking. It is reasonable that consumers (or consumer organisations on their behalf) have access to information about the history of a product. A relevant question is: How water intensive is a particular product that is for sale and to which extent does it relate to water depletion and/or pollution? Establishing a mechanism that makes sure that such information is available is not an easy task. It requires a form of accounting along production and supply chains that accumulates relevant information all the way to the end point of a chain.

Governments that put interest in 'sustainable consumption' should translate this interest into their trade policy. The European Union, given the fact that about 40 % of the total water footprint of the EU citizens lies outside its own territory (Mekonnen and Hoekstra 2011), may strive towards more transparency about the water impacts of imported products. Achieving such a goal will obviously be much easier if there is international cooperation in this field. In cases where industrialised countries import feed from developing countries, the former can support the latter within the context of development cooperation policy in reducing the impacts on local water systems by helping to set up better systems of water governance.

Business can have a key role as well, particularly the large food processors and retailers. Since they form an intermediary between farmers and consumers, they are the ones that have to pass on key information about the products that they are trading. As big customers, they can also put pressure on and support farmers to actually reduce their water footprint and require them to provide proper environmental accounts. If it comes to water accounting, there are currently several parallel processes going on in the business world. First of all, there is an increasing interest in the water use in supply chains, on top of the traditional interest in their own operational water use. Second, several companies, including for instance Unilever and The Coca Cola Company, have started to explore how water footprint accounting can be practically implemented. Some businesses think about extending their annual environmental report with a paragraph on the water footprint of their business. Others speak about water labelling of products (either on the product itself or through information available online), and yet others explore the idea of water certification for companies. The interest in water footprint accounting comes from various business sectors, ranging from the food and beverage industry to the apparel and paper industry.

9 Conclusion

The interest in the water footprint in the food sector is growing rapidly, but most interest thus far comes from the beverage sector (Sarni 2011). Besides, most companies still restrict their interest in water to their own operational water footprint, leaving the supply chain water footprint out of scope. Little interest in water has been shown in the meat and dairy sectors, which is surprising given the fact that the meat and dairy sectors contribute more than a quarter to the global water footprint of humanity. Also from the governmental side, there is hardly any attention to the relation between animal products and water resources. There does not exist a national water plan in the world that addresses the issue that meat and dairy are among the most water-intensive consumer products, let alone that national water policies somehow involve consumers or the meat and dairy industry in this respect. Water policies are often focussed at 'sustainable production', but they seldom address 'sustainable consumption'. They address the issue of water-use efficiency within agriculture (more crop per drop), but hardly ever the issue of water-use efficiency in the food system as a whole (more kcal per drop). The advantage of involving the whole supply chain is that enormous leverage can be created to establish change.

The issue of wise water governance is a shared responsibility of consumers, governments, businesses and investors. Each of those players has a different role. Consumers should demand transparency about the water consumption and pollution underlying consumer products from business and governments, so that one is better informed about associated water resource use and impacts. Consumers can choose to consume less animal products or choose, whenever proper information allows, for products with a water footprint that meets a certain benchmark. National governments can—preferably in the context of an international agreement—put regulations in place that urge businesses along the supply chain to cooperate in creating product transparency. Governments can also tune their trade and development cooperation policies towards their wish to promote consumption of and trade in sustainable products. Governments should further play a leading role in establishing water footprint ceilings per river basin, to ensure that in each river basin, the water footprint does not exceed available water resources. Companies, particularly big food processors and retailers, can use their power in the supply chain to effectuate product transparency. They can also cooperate in water labelling, certification and benchmarking schemes and produce annual water accounts that include a report of the supply chain water footprints and associated impacts of their products. Investors, finally, can be an important driving force to encourage companies to put water risk and good water stewardship higher on their corporate agenda. Some steps in creating product transparency in the food sector have been made to address concerns of product quality and public health. It is likely that in the future, there will be increasing interest in transparency regarding environmental issues like water resource use as well.

References

Aldaya, M. M., & Hoekstra, A. Y. (2010). The water needed for Italians to eat pasta and pizza. *Agricultural Systems, 103*, 351–360.

Allan, T. (2011). *Virtual water: Tackling the threat to our planet's most precious resource*. Taurus, London, UK: I.B.

Ercin, A. E., Aldaya, M. M., & Hoekstra, A. Y. (2011). *The water footprint of soy milk and soy burger and equivalent animal products*. Value of Water Research Report Series No. 49. Delft, The Netherlands: UNESCO-IHE.

FAO. (2005). *Livestock policy brief 02*. Rome, Italy: Food and Agriculture Organization.

FAO. (2011). *Food balance sheets*. Rome, Italy: FAOSTAT, Food and Agriculture Organization.

Galloway, J. N., Burke, M., Bradford, G. E., Naylor, R., Falcon, W., Chapagain, A. K., et al. (2007). International trade in meat: The tip of the pork chop. *Ambio, 36*, 622–629.

Gerbens-Leenes, P. W., Mekonnen, M. M., & Hoekstra, A. Y. (2011). *A comparative study on the water footprint of poultry, pork and beef in different countries and production systems*. Value of Water Research Report Series No. 55. Delft, the Netherlands: UNESCO-IHE.

Hendy, C. R. C., Kleih, U., Crawshaw, R., & Phillips, M. (1995). *Livestock and the environment finding a balance: Interactions between livestock production systems and the environment, impact domain: Concentrate feed demand*. Rome, Italy: Food and Agriculture Organization.

Hoekstra, A. Y. (2012). The hidden water resource use behind meat and dairy. *Animal Frontiers, 2* (2), 3–8.

Hoekstra, A. Y. (2013). *The water footprint of modern consumer society*. London, UK: Routledge.

Hoekstra, A. Y., & Chapagain, A. K. (2008). *Globalization of water: Sharing the planet's freshwater resources*. Oxford, UK: Blackwell Publishing.

Hoekstra, A. Y., Chapagain, A. K., Aldaya, M. M., & Mekonnen, M. M. (2011). *The water footprint assessment manual: Setting the global standard*. London, UK: Earthscan.

Hoekstra, A. Y., & Mekonnen, M. M. (2012). The water footprint of humanity. *Proceedings of the National Academy of Sciences, 109*(9), 3232–3237.

McGuire, V. L. (2007). *Water-level changes in the High Plains Aquifer, predevelopment to 2005 and 2003 to 2005*. U.S. Geological Survey Scientific Investigations Report 2006–5324.

Mekonnen, M. M., & Hoekstra, A. Y. (2011). *National water footprint accounts: The green, blue and grey water footprint of production and consumption*. Value of Water Research Report Series No. 50. Delft, The Netherlands: UNESCO-IHE.

Mekonnen, M. M., & Hoekstra, A. Y. (2012). A global assessment of the water footprint of farm animal products. *Ecosystems, 15*(3), 401–415.

Naylor, R., Steinfeld, H., Falcon, W., Galloway, J., Smil, V., Bradford, E., et al. (2005). Agriculture: Losing the links between livestock and land. *Science, 310*, 1621–1622.

Pittock, J., & Connell, D. (2010). Australia demonstrates the planet's future: Water and climate in the Murray-Darling Basin. *International Journal of Water Resources Development, 26*, 561–578.

Sarni, W. (2011). *Corporate water strategies*. London, UK: Earthscan.

Steinfeld, H., Gerber, P., Wassenaar, T., Castel, V., Rosales, M., & de Haan, C. (2006). *Livestock's long shadow: Environmental issues and options*. Rome, Italy: Food and Agriculture Organization.

Part II
Virtual Water, Humans and the Environment

Water Resources in the Anthropocene Age: A Planetary Urgency

Eva Alessi and Gianfranco Bologna

Global environmental change (GEC) is the global change that human activity is causing in the natural systems of our wonderful planet. The international scientific community which is involved in GEC has been calling on, for years, the political and economic world to take action so that our society and our development models may finally start their journey on the road to global sustainability.

In March 2012, London hosted the latest major scientific conference on global change, "Planet under pressure. New knowledge towards solutions" (www.planetunderpresssure2012.net), organized by the leading international scientific organization International Council for Science (ICSU www.icsu.org), and by the Earth System Science Partnership (ESSP, www.essp.org, now transformed in www.futureearth.org), the worldwide partnership that brings together the most authoritative research programmes on global change. About 3000 scientists, academics, and government experts gathered to discuss global change, focusing on the state of the planet's natural systems and on the knowledge we have about the strain caused to them by human activities. Operative proposals were presented that could be urgently implemented to change the direction of the present model of socio-economic development which, up to now, has clearly proved unsustainable for the immediate future. The conference concluded with the statement of a very straightforward "State of the Planet Declaration," clearly outlining the state of health of our natural systems. The declaration was meant to remind us that the research on global change proves that the complex planetary system which, over the centuries, has sustained human well-being and civilizations, is at risk. Without adopting urgent measures, we will have increasing difficulties in facing the threats posed to our vital resources, among them water, which are at the basis of our alimentation and biodiversity. These threats risk to intensify the economic, ecological, and social crises, creating the potential for a global humanitarian emergency. The world's existing interconnected and

E. Alessi (✉) · G. Bologna
WWF, Rome, Italy
e-mail: e.alessi@wwf.it

G. Bologna
e-mail: g.bologna@wwf.it

© Springer International Publishing Switzerland 2015

M. Antonelli and F. Greco (eds.), *The Water We Eat*,
Springer Water, DOI 10.1007/978-3-319-16393-2_4

interdependent economic, social, cultural, and political systems are placing a very strong pressure on the natural systems, causing fundamental changes to the Earth as a system and leading human societies beyond the natural "planetary boundaries" within they should act instead.

However, the same interconnection can be used to identify the solutions to these serious problems. Global sustainability must become the basis of our societies.

The declaration recalls that the human impact on the Earth's system can now be compared to the great geological planetary changes, such as those occurring during the ice ages. Consensus has grown on the fact that our planet has entered a new geological era called the "Anthropocene" (originally proposed in 2000 by the Nobel Prize Winner for Chemistry, Paul Crutzen, Crutzen, 2002). In this new era, many processes belonging to the Earth system and to the living "factory" of its ecosystems are now dominated by human activities. Large-scale abrupt changes, identified by the research on past environmental changes, indicate that these sudden changes could also take place in our immediate future. This knowledge has encouraged researchers to try identifying the thresholds (the "tipping points," the critical points, or threshold effects) and the planetary and regional boundaries that, once crossed, could generate environmental and social changes ultimately ungovernable by our societies.

The declaration calls for a fundamental reorientation and reorganization of our international and national institutions to move toward an effective Earth system governance, a concrete commitment to a proposal for sustainable development objectives, such as those for global sustainability, and the recognition of monetary and non-monetary values of common goods, such as ecosystem services and large-scale planetary environments, the oceans, and the atmosphere.

During the conference, the international scientific community dedicated to studies on global change announced the new research program for the next ten years called "Future Earth: Research for Global Sustainability" (www.icsu.org/future-earth, and www.futureearth.org).

The vulnerability of the Earth system to human activities also fully demonstrates the intrinsic vulnerability of our civilization, vis a vis what we have caused to the natural systems and which, inevitably, will have an effect on our well-being and survival. It is now time to change direction and to also modify our established concept of security.

The noted environmentalist, Lester Brown, wrote in his book *Plan B 4.0* (Brown 2010): "The situation in which we find ourselves pushes us to redefine security in twenty-first century terms. The time when military forces were the prime threat to security has faded into the past. The threats now are climate volatility, spreading water shortages, continuing population growth, spreading hunger and failing states. The challenge is to devise new fiscal priorities that match these new security threats. We are facing issues of near-overwhelming complexity and unprecedented urgency. … We are in a race between natural and political tipping points, but we do not know exactly where nature's tipping points are. Nature determines these. Nature is the timekeeper, but we cannot see the clock. The notion that our civilization is approaching its demise is not an easy concept to grasp or accept. It is difficult to imagine something we have not previously experienced. We hardly have even the vocabulary, much less the experience, to discuss this prospect… Since it is the destruction of the economy's natural supports and disruption of the climatic system that are driving the world to the edge, these are the trends that must be reversed. To do so requires extraordinarily demanding measures, a fast shift away from business as usual… One thing is certain—we are facing greater change than any generation in history. What is not clear is the source of this change. Will we stay with business as usual and enter a period of economic decline and spreading chaos? Or will we quickly reorder priorities, acting at wartime speed to move the world onto an economic path that can sustain civilization?"

1 Water Demand in the Anthropocene

Freshwater is an integral part of the Earth system "machine" and a very important key to understanding the magnitude of this global change. The heating of the Earth's surface due to the increase in greenhouse gases, caused by humans, also changes the water cycles. Many other anthropic factors—the widespread change in land use, the engineering of river beds, irrigation and other water uses, the disappearance of aquatic habitats, pollution—also influence the global water system.

Consequently, there follows the decline in biodiversity and the deterioration of the ecosystems and their resources: Many species risk extinction, while the rainforests, the coral reefs, and the wet zones suffer the devastating effects of human activities. Extreme phenomena, such as flooding and droughts, have increased in intensity and frequency, and the shortages in food and water have become a serious concern for the entire world.

The history of the constant evolution of our planet began about 4.6 billion years ago and has witnessed a series of large-scale geological and biological upheavals. The eras and epochs have reflected these events, taking place on a geological time scale, each marked by extraordinary occurrences that have changed the planet's history, such as the ice ages, the plate tectonics, or the mass extinction of living species. A recent event occurred approximately 11,500 years ago when the retreat of the glaciers made way for the current interglacial period, the Holocene, creating most of the present geological formations, as for example, the potentially arable land, the river deltas, and fluvial deposits. Over the last 10,000 years, our species has been able to enjoy a situation, even with the evolutionary dynamics of the natural systems, of a modest climate and environmental stability which has allowed us to grow and now reach a population of 7 billion.

In the more recent phase of the Holocene, human pressure on the natural systems has become so heavy that it can be compared with the large-scale geological forces that have modified the planet throughout its entire life. From the second half of the twentieth century, the history of the human species has witnessed another important shift: More than a half of the rural population has become urban, supported during the green revolution by an industrial agriculture based on the use of fertilizers and pesticides, intensive irrigation, mechanization, and widespread use of fossil fuels. However, anthropic activities have accelerated and/or intensified many important natural processes, for example, soil erosion, biogeochemical cycles of carbon, nitrogen and phosphorus, the increase in the greenhouse effect, and they have slowed down others, as for example, the quantity of water and sediments that reach the sea via rivers. Although very recent and rapid, as we have already mentioned, this period may represent the beginning of a new geological epoch, the Anthropocene, characterized by the rapid change in the human–ecosystem relationship, and to the extent that humanity could be considered as a global force in itself (Global Water System Project 2012).

The term Anthropocene was introduced to actually underline the growing predominance of humanity in determining GECs and is characterized, among other things, by the growing demand for water and the decrease in its availability. Water use has significantly increased since the advent of the industrial revolution and still even more rapidly over the last four decades. This has caused global changes in river flows (Shiklomanov and Rodda 2003), spatial patterns, and seasonal times for "global water vapor flows" (Gordon et al. 2005). It is estimated that because of water withdrawals, 25 % of the world's river basins dry up before they reach the oceans (Molden et al. 2007).

For example, in the 1990s, the Yellow River in China dried up for long periods of time along its course and at its estuary; huge problems occurred in attempting to conserve the flow of the Murray River in Australia and the Rio Grande on the border between Mexico and the United States; both rivers now experience long periods of drought. In order to meet the ever increasing demand, water is also transported over long distances from one river to another, which can aggravate ecological impacts. Sometimes, this happens on a large scale, as is the case in the "South–North Water Transfer Project" in China (this project plans, by 2050, to divert 44.8 billion cubic meters of water per year from the Yangtze River in the south of China to the Yellow River basin in the north, *ndr*). The reasons underlying this enormous water use (for agricultural, domestic, industrial and energy production purposes) include population growth, economic development, globalization, changes in consumption models and, consequently, in dieting models. An example of the scale of total global anthropic water use lies in plant production that, cumulatively, amounts to about 7000 km^3/year of water. If we include permanent grazing lands, the total evapotranspiration for food production amounts to 15–20,000 km^3/year (Hoff 2009). In order to understand the magnitude of these numbers, this amount corresponds to around the half of the total superficial water discharge of all the rivers on Earth (Hoff 2009). Given the present productivity levels and access to food, a further 5000 km^3/year will be needed to feed the world's population by 2050.

The growing water demand is causing a decrease in quantity, quality, and regularity of the water available for our ecosystems. This, in turn, is resulting in the loss and degradation of ecosystems' biodiversity, which, consequently, diminishes the capacity of those ecosystems to provide the essential services which are necessary to support our societies and life on Earth.

Water scarcity, which can generally be understood as the lack of access to adequate quantities of water for human and environmental use, is acknowledged in many countries as a serious and growing problem. Consequently, "water scarcity" is the term regularly used by the media, in government reports, by NGOs and by international organizations such as the United Nations and the OECD, as well as in academic literature, to highlight the areas where water resources have been subjected to pressure or stress. Therefore, the scientific community is increasingly reminding us of the urgency to act and the need to rapidly intervene to change the direction of our development models that are based on a material and quantitative economic growth. An internationally famous team of scientists, led by Johan Rockström, director of the Stockholm Resilience Center, identified in 2009 the nine thresholds of planetary boundaries (linked to established ecological parameters) which human activities should not step over in order to avoid serious biosphere imbalances. The same scientists confirmed, as well, that in many situations we are already near the "threshold." As evidence increases regarding how, on a global scale, we are facing a water crisis, the global use of freshwater is, obviously, one of the nine planetary boundaries identified (Rockström et al. 2009a, b; Steffen et al. 2015).

The alteration in global water cycles affects biodiversity, food, the health, and the ecological functionality of the ecosystems. For example, changing freshwater habitats may have deep repercussions on fish stocks, on the level of carbon capture and storage, on climate regulation, undermining the resilience of land and water ecosystems.

The deterioration of the global water resources threatens human survival as it determines:

- the loss of soil moisture (also defined as "green" water, being the basis of non-irrigated agriculture), caused by land degradation and deforestation which threatens primary production, carbon capture, and storage;
- exploitation and changes in surface water runoffs (defined as "blue" water), which threatens agricultural and domestic needs;
- the impact on climate regulation caused by the decline in "global water vapor flows" which change local and regional precipitation patterns.

Estimates indicate that 90 % of the world's green water flow is necessary to support vital ecosystem services (Rockström et al. 1999), while 20–50 % of the average annual blue water flow in the catchment basins is necessary to support aquatic ecosystems (Smakhtin 2008).

The inexorable increase in water demand to produce food, to supply industries, urban and rural populations, has resulted in an increasing freshwater scarcity in many parts of the world. In fact, while the erosion of the Earth's productive surface stratum began with the first farming of wheat and barley, the trend in lowering the groundwater levels is historically more recent, since the technology required to pump water from underground aquifers is only a few decades old. The wells gradually dry up and the depletion of groundwater aquifers occurs where there is the habit to pump excessive quantities of water. An increasing number of rivers are drying up for long periods before reaching the sea. In many areas, groundwater is pumped at rates that exceed the groundwater replenishment capacity, with the result that the groundwater levels are rapidly dropping in at least 20 countries, including India, China, and the United States. The three most populated countries in the world where half of the world's cereal crops are grown (Brown 2011). Consequently, it is relatively clear how water shortage can turn into food shortage, taking into consideration the data that show how, for example, 40 % of the world's wheat harvest originates from irrigated lands. A World Bank study shows that 175 million people in India are fed thanks to over-exploited groundwater, while in China we are speaking of more than 130 million people (Brown 2004).

As Werner Aeschbach-Hertig and Gleeson (2012) recall, humanity, today, also intensively uses so-called fossil water (the underground water accumulated throughout the Earth's geological history, scarcely conditioned by the surface water cycles and, therefore, with a very slow renewability rate). As these academics demonstrated, there is a widespread depletion of fossil water reservoirs, particularly in regions of high fossil water use for irrigation purposes in China, India, and the United States. The situation is especially serious as it does not only concern the withdrawal-renewability deficit from surface water. The rapid increase in evapotranspiration should be also taken into account, which is augmented by climate change, growing anthropic pressures and by the various impacts on agricultural activities.

Climatic changes also threaten food security. There is a threshold where the increase in temperatures represents a problem for agricultural production. Every increase of 1 during the growing season can imply, for farmers, a decrease of 10 % in wheat, rice, and maize harvests. From 1970 until today, the planet's average surface temperature has increased by more than 0.7 °C (IPCC 2007). Increasingly more often, governments, societies, and communities have become worried about the future availability and sustainability of water resources. Over the last 20 years, researchers have developed a number of parameters and indicators to describe and map the geography of global water scarcity. These include, for example, the ratio between the human population and the renewable water systems (Falkenmark 1989), the ratio between water withdrawals and renewable supply, and they contribute to documenting the spread of water scarcity over time.

Meeting the demand of a world population that is growing every year by about 80 million people has become increasingly more difficult. If the reduction in food consumption, driven by the present serious economic crisis, is a new situation for many countries in the world, for many others further sacrifices are not possible. Food is the "Achille's heel" of our societies and risks becoming an important factor of instability, also politically. To avoid a collapse, also from an environmental viewpoint, societies as a whole need to mobilize. No one can afford to be merely a spectator any more (Brown 2012).

2 Water Resource Management

Water resources and their management, up to today, have been approached on a local, or even catchment basin level. Only recently has it been recognized that water resources are, at the same time, subject to and an integral part of global change and globalization. The idea of a global water system with interdependent social and ecological elements has only been recently established (Alcamo et al. 2008).

In 2010, in the scientific journal "Nature" (Vorosmarty et al. 2010) a very interesting study was presented which documents how almost 80 % of the human population, or 5.6 billion people (out of the 7 billion that now live on our planet), live in areas where there exists a high risk concerning water supply security and the state of health of the biodiversity of freshwater habitats. This involves ecosystems highly threatened by pollution, dam construction, the presence of invasive species, coastal habitat transformation, etc. The study is the first which correlates the factors threatening water supply security for humans to the state of health of the biodiversity of the ecosystems which provide water. It clearly shows the need and urgency for a careful and coordinated management of the water demand of human society, preserving and guaranteeing the services which the freshwater ecosystems provide for our well-being and survival. The stress factors that threaten freshwater ecosystems jeopardize the security of water reserves for human use and 65 % of the world's river habitats are now classified at risk—from moderate to very high. This affects the survival of thousands of aquatic species. The main rivers under threat are in India, Europe, the Middle East, Southeast Asian countries, and the United States. Only a small fraction of the world's rivers appear to be not significantly compromised by human activities, as can be seen, for example, in some remote areas of the Amazon River and the Congo Basin.

The WWF, in agreement with the international scientific community, has stressed for years how it is no longer possible to refer to human water security as something unconnected to the value of ecosystems' biodiversity, from which it originates. Practically almost all of humanity lives near a water source, either at the end of a water pipe or in proximity of a river. We need water to survive, to grow crops, to generate electricity, and to produce the goods for our daily use. Though less than 1 % of the world's water resources is available, that quantity must satisfy both human and environmental needs, the two being inseparable. The key issue, therefore, is to guarantee sufficient water of adequate quality to the humanity, preventing the destruction of the basic ecosystems necessary for its supply, such as rivers, lakes, and groundwater.

Today, the services provided to the economy and human society by freshwater ecosystems, including the supply of water, are exploited well over their sustainable levels. This was clearly shown by the "Millennium Ecosystem Assessment," the international study published in 2005 and sponsored by the UNO, on the state of health of the planet's ecosystems and their future (www.maweb.org).

Furthermore, it is forecasted that water resource demand, the so-called human water footprint (www.waterfootprint.org), is continuing to increase in many parts of the world. The main impacts of the human water footprint on the freshwater ecosystems derive from the increase in river fragmentation, in excessive water withdrawal, and from pollution. In addition to that, the looming impacts of current climate change could exacerbate the situation (WWF 2012).

The chain reaction on the global scale of water scarcity was fully understood from the moment the methods of water footprint calculation highlighted how much countries and economies depend on the trade in virtual water embedded in all our food and services (from a cup of coffee to a cotton T-shirt, or a steak), but never taken into account in any economic evaluation.

The growing demand for water and hydro-electric energy, along with the attempts to control flooding and to improve river navigation, have led to the construction of dams and other infrastructures, such as locks, underwater dams, and levees along most of the world's large rivers. In total, out of the 177 rivers of more than 1000 km in length, only 64 flow freely without being hindered by dams or other barriers, as shown in a WWF study published in 2006 ("Free-flowing rivers – Economic luxury or ecological necessity?", www.panda.org). A water infrastructure can result in benefits, but it can also deeply impact the freshwater ecosystems and populations that depend on the services provided by those ecosystems. Dams alter river flows, changing the quantity, the times, and the quality of the water that flows downstream. Moreover, larger dams can totally interrupt the ecological linkages between the upstream and downstream habitats, creating serious problems, for example, to migratory fish species.

The most recent research calculates that dam construction has a negative impact on the life and existence of about 500 million people.

The economic and financial crisis is strongly hitting many countries worldwide. The climate change and ecosystem crisis is doing exactly the same with even more serious repercussions as it is incredibly difficult to rebalance the ecological deficit which we have caused. The consequences of climate change and of the overexploitation of the natural resources, such as forests, agricultural land, fish species, and freshwater, will impact differently in different parts of the world. Some regions will suffer more than others, but in the long term no area of the world will be able to protect itself or avoid these negative consequences. Ultimately, energy food and water suppliers will be those most at risk. Similar to the financial crisis, which had been forecasted for many years beforehand, but had gone unheeded, the same is happening for climate change and environmental deterioration. Why then, despite the many warnings, has little or nothing been done? The environment is now in a much worse state compared to 50 years ago, its evolution being dominated by human activities. This is occurring notwithstanding the increasingly detailed and numerous studies that have clearly shown how our societies are on a collision course with nature. Environmental issues regularly end up at the top of the list of priority questions in opinion polls, but any real action is still a long way off. Therefore, it has become increasingly important to establish economies based on the efficient use of our resources, through a tax system that transfers the revenues of current taxation on employment into strategies toward a sound utilization of resources and sound pollution control, and on clear guidelines and objectives for global sustainability.

Adopting the family of "footprint" indicators (ecological footprint, carbon footprint, water footprint, material footprint, etc.) could be very useful in contributing to better transparency for consumers. However, above all, there is an extreme necessity to act and to change the direction of our social and economic development models. Inaction does nothing but aggravate the solution to the problems.

References

Aeschbach-Hertig, W., & Gleeson, T. (2012). Regional strategies for accelerating the global problem of groundwater depletion. *Nature Geoscience, 5*, 853–861.

Alcamo, J., Vorosmarty, C. J., Naiman, R., et al. (2008). A grand challenge for fresh water research: Understanding the global water system. *Environmental Research Letters, 3*(1), 010202.

Brown, L. R. (2004). *Outgrowing the earth*. New York: W.W. Norton & Company.

Brown, L. R. (2010), *Piano B 4.0*, It. ed. edited by G. Bologna, Edizioni Ambiente.

Brown, L. R. (2011), *Un mondo al bivio:come prevenire il collasso ambientale ed economico*, It. ed. Edited by G. Bologna, Edizioni Ambiente.

Brown, L. R. (2012), *9 miliardi di posti a tavola*, It. ed. Edited by G. Bologna, Edizioni Ambiente.

Crutzen, P. J. (2002). Geology of mankind. *Nature, 415*, n.23.

Falkenmark, M. (1989). The massive water scarcity now threatening Africa: Why isn't it being addressed? *Ambio, 18*(2), 112–118.

Global Water System Project (2012). Water in the Anthropocene. *Global water news*, n. 12, October 2012. www.gwsp.org.

Gordon, L., et al. (2005). Human modification of global water vapor flows from the land surface. *Proceedings of the National Academy of Sciences, 102*(21), 7612–7617.

Hoff, H. (2009). Global water resources and their management. *Current Opinion in Environmental Sustainability, 1*(2), 141–147.

IPCC (Intergovernmental Panel on Climate Change). (2007), *Summary for policymakers*. Climate change 2007: The physical science basis. Contribution of working group 1 to the fourth assessment report of the intergovernmental panel on climate change. Cambridge: Cambridge University Press.

Molden, D., et al. (2007), Trends in water and agricultural development. In D. Molden (Ed.), IWMI, *Water for food, water for life: A comprehensive assessment of water management in agriculture*. London: Earthscan.

Rockström, J., et al. (1999). Linkages among water vapour flows, food production, and terrestrial ecosystem services. *Ecology and society, 3*, n. 2, art. 5.

Rockström, J., et al. (2009a). A Safe operating space for humanity. *Nature, 461*, 472–475.

Rockström, J., et al. (2009b) Planetary boundaries: Exploring the safe operating space for humanity. *Ecology and Society, 14*, n. 2, art. 32. www.ecologyandsociety.org/vol14/iss2/art32.

Shiklomanov, I. A., & Rodda, J. C. (2003). *World water resources at the beginning of the 21st century, UNESCO*. Cambridge: Cambridge University Press.

Smakhtin, V. (2008). Basin closure and environmental flow requirements. *International Journal of Water Resources Development, 24*(2), 227–233.

Steffen, W., et al. (2015). Planetary boundaries: Guiding human development on a changing planet. *Science*, January 2015. doi:10.1126/Science.1259855.

Vorosmarty, C. J., et al. (2010). Global threats to human water security and river biodiversity. *Nature, 467*, 555–561.

WWF (2012). *Living Planet Report 2012*, in collaboration with the Global Footprint Network, Zoological Society of London, European Space Agency. www.panda.org.

Water in Food

Lynne Chatterton

The terms virtual water and food security are increasingly used to enable us to focus on the need to conserve resources that pertain to the growing and eating of food. By understanding what water is required for food after the growing process, the complete role of water in food is revealed. That there is a looming crisis in the availability of food to many consumers is not denied. But as world famous economist Amartya Sen has written, there is sufficient food in the world today to feed everyone, and the problem is in the inequalities of distribution.

What is a real crisis is the loss of arable land to industry, housing and the agribusinesses that produce profit but not the nutritious food that we need to remain healthy.

Slopes cleared of trees and undergrowth for these purposes become a channel for erosion and the cause of destructive flooding when heavy rainfalls. The bitumening of roads prevents rainfall from entering the soil where it will nurture vegetation and trees and provide food for the animals we eat. Intense city conurbations such as London frustrate rainfall from being absorbed and used in the production of food, yet they create increasingly unsustainable demands for domestic and drinking water. Many of the industries operating within cities also use large quantities of water. Gardens, historically a source of food for families, are either non-existent, minute or devoted, as public parks are, to aesthetically pleasing trees, shrubs and flowers. Rainwater tanks to harvest rain from roofs are rare and only found on allotments, where a very small but increasing number of people are growing their own food.

Another growing crisis is the wasteful use of water in the food chain. We can see this crisis in action when we examine the food available to most consumers in the Anglo-Saxon world.

While we profess to be enamoured of "ethnic" food, we have not yet understood that it is based on fresh, largely untreated fruit and vegetables available from daily markets or domestic small holdings. The meat, from animals fed on pasture and grain in stubble, is sold the same day. Much of the ethnic food that 87 % of people

L. Chatterton (✉)
Podere valle Pulcini, Castel di Fiori, 05010 Montegabbione, TR, Italy
e-mail: pulcinipress@hotmail.com

© Springer International Publishing Switzerland 2015

M. Antonelli and F. Greco (eds.), *The Water We Eat*,
Springer Water, DOI 10.1007/978-3-319-16393-2_5

buy and consume from supermarkets is not "fresh" and is only a pale shadow of the real thing. It is also deficient in nutrition compared to fresh food, eaten raw or cooked daily in the home.

In spite of manufactured food becoming available in cities within the so-called developing world, the majority of the population still demands and eats food cooked from fresh meat and vegetables grown sufficiently close enough to town and city markets to avoid the deterioration that lowers nutrition.

This is, of course, a generalisation. Some food does come from animals force fed in feedlots and industrial sheds—a system that has been heavily promoted in many developing countries by development agencies—and by the agribusinesses, based on plastic tunnels, chemical stimulation and control of pests by pesticides. These are used to produce vegetables mainly for export to supermarkets in the West.

The determination of Western style supermarkets to expand into countries such as India poses a threat also to the way food is grown and sold to the consumer in Asia, many Arab countries, Turkey and the Southern Mediterranean. Such an invasion, if achieved, would result in food coming by and large from manufactured products requiring much larger quantities of water than the current food chain that consumers rely on. This is fresh food, grown under rainfall or managed irrigation using water stored in domestic tanks or large reserves, and in some cases direct from rivers. Fresh food grown in domestic plots, and on small farms, is a remarkably economical use of water to produce food. Water that supplements rainfall is stored and used for dry periods. When the food chain from farm to consumer is direct and daily, it cuts out the manufacturing process that transforms fresh food into product.

1 The Food Chain

If we take a plate of typical food from Anglo-Saxon sources, we will see that much of it is grown in other countries, processed and distributed through a long and complicated food chain.

If we take a plate of typical food from the countries named above, we will see that much of it is grown very close to where the consumer lives, which is largely bought and consumed daily.

If we follow the operations needed to transform food from plant and animal into consumer product, we can see how much water is used over and above that required to grow the food to its natural maturity.

Meat provides a good example of this.

Meat from animals fed on grain, sometimes grown as a fodder crop under irrigation and concentrates manufactured in factories, will consume much more water before it reaches our tables than animals fed on pasture in summer and hay and silage during winter. The demands of water for irrigated wheat grown for fodder, the processing of concentrates, the packing, transport, and other operations required to place meat on supermarket shelves all consume water as they take their place in the meat food chain. The hormones administered to animals on this regime to stimulate growth require water in their manufacture as do the antibiotics and other prophylactics used widely to protect animals from illnesses common to intensive production.

To turn the animal into meat from this source, it is necessary to first slaughter and prioritise parts of the carcase. Some of this meat is turned into processed meat sold in supermarkets and acquired by many commercial food catering outlets. Much of it is used in the precooked "instant" meals packs that have become so ubiquitous over the last 30 or 40 years. During the processing operation of this product, much water is used—some of it injected into the carcase to cause the meat to swell and appear larger than it is thus increasing the profit to the eventual retailer. When this meat is cut into small portions, and some of it is minced for hamburgers, more water is consumed. The factory processes required to turn meat into profitable product are water intensive, the packaging for the display of this meat in shops requires water in its production, and the need for strict date stamping due to the perishability of meat leads to waste as consumers reject some packages that the retailer must then discard or are kept too long in the household refrigerator or freezer and must be discarded for fear of food poisoning. Refrigerators and freezers necessary for the storage of meat products also require water to build and function—thus making yet another demand on the water resource.

Animals fed on pasture and silage only consume water that comes from rainfall. The major use of water for these animals occurs during the abattoir process. They are not fed hormones and only need antibiotics occasionally for specific maladies and rare infections.

Fresh meat sold to consumers daily curtails the food corporation drive for profits. It deprives the packaging, transport and storage industries of markets, although the advertising industry remains untouched. The producer of the meat, the farmer and the farmer's family, receives better prices for their animals, and the consumer benefits from the increased nutrition available in fresh meat grown in concert with nature. Much less water is used in producing food to the consumer and this benefits us all.

As with all matters pertaining to food and farming, it is unwise to draw too many conclusions from broad-based and averaged out statistics. Recent advice to stop eating meat and thus save water is an example of this. The water content quoted appears to have been derived from meat from animals intensively raised—what is commonly known as factory farming. Thus, an animal raised on a factory farm, fed grain and concentrates reliant on irrigation, will certainly require a lot of water. As is shown above, the water waste does not stop there—the food chain required for the marketing of this meat is heavily dependent on water to operate the various processes needed to transform the meat into product for the consumer.

The same factors apply to poultry. Poultry raised in the open air and allowed to graze green fodder that is supplemented with grain, so-called free range, are healthier and less demanding of water than intensive shedding based on industrial feed, growth hormones and antibiotics.

Dairy products from cows fed on pasture and allowed to follow natural cycles will require much less water than factory farmed dairy cows, bred for maximum milk production, to which are added antibiotics and growth compounds.

Thus, meat from a lamb from, say, the Welsh hills or from Australia, where rainfed pasture is the principal source of feed, will utilise much less water than meat from a lamb, grown in sheds or feedlots, fed on grain and concentrates to which are added hormones and prophylactic substances.

It goes without saying that meat sold to the consumer will be more nourishing and healthy if it comes from grazed and healthy animals, and if processing, packaging, storage and transport can be kept to a minimum.

2 Wheat and Barley

Most of the wheat grown in the world is grown on rainfall alone. Wheat, grown under irrigation, costs much more to produce and requires much water to mature and yield a crop.

It will be said that irrigation provides more security against seasonal extremes, but if there is a drought, then irrigation will become problematical. Water in rivers and dams also suffers from drought. Monocultural crops such as wheat and barley are grown in open fields and require a lot of water so drip irrigation is impractical and large centre pivot sprinklers are extravagant in their water use, losing much of it through dehydration before it is available to the plants. The crop will also receive what rainfall there is as the sky cannot be turned off.

Bread and other food obtained from wheat and barley grown under rainfall alone is much less costly in terms of water used than that obtained from irrigated grain.

There is another distinction between wheat grown as a monocultural crop and wheat and barley grown in rotation with self-regenerating pasture such as medicago. Wheat grown without this rotation requires large quantities of nitrogen fertiliser to yield productively, and water is needed in the process of manufacturing this.

A pasture/wheat rotation requires one-fifth of the cultivation needed compared to wheat reliant on nitrogen fertiliser. The water requirements of each passage of the machinery are not insubstantial, if somewhat hidden from the casual observer.

Yet again, milling, processing, packing, storage, etc., all require water over and above the needs of the growing plant.

3 Olives

Olives and olive oil are staple food requirements in the WANA region, and countries in Southern Europe and Turkey.

They grow on rainfall alone and are often overlooked as an important source of food in these regions. Yet they are one of the most water economical foods of all.

Olive trees thrive on hot, dry conditions in summer and low winter rainfall.

The adoption of olive cultivation in Australia, New Zealand and parts of North America has seen the use of additional water applied in irrigation systems to trees in dry summers to induce rapid growth of the young trees and also to encourage higher than traditional yields. This is a commercial decision and not a necessity. The only extra water required for the oil to be available is that used to wash the fruit as it begins the process of oil being extracted. Once the olives have been crushed, the crushed material moves to the oil separation phase which takes place under centrifuge using the olive juice obtained from the crushing process to separate out the oil. Nothing further is added—the only water used at this stage is to wash the containers before the oil is poured into them.

Olive fruit for the table requires only the saline water needed to leach out the bitterness before they are packed in jars and preserved in either oil or salt.

Commercial table olives in some countries are soaked in weak caustic soda solution to remove the bitterness and then must be washed in large amounts of water to remove the caustic soda.

If they are sold in supermarkets, they are usually vacuum packed in plastic pouches or made available in tins. Water to manufacture plastic and tins is necessary for both these forms of preservation. And, again, the exigencies of the large supermarket use even more water.

4 Fruit and Vegetables

Vegetables, in particular, require ample water if they are to produce healthy and nutritious food.

In Southern Europe and parts of the WANA region and Turkey, some vegetable crops are grown on moisture in the soil from the winter rainfall—for example melons and early tomatoes.

There are vegetables that grow and are harvested during various phases in winter such as cabbage, cauliflower, onions and garlic. These do not require irrigation, unless the climate is particularly arid.

In Europe and North America, the summer season for vegetables is relatively short and growth must begin early and be maintained until the plant reaches maturity. Almost daily watering is needed as any short period of dryness will check growth and affect the plant's productivity.

However, the way in which the water is applied is important in understanding the amount of water needed. Open-field crops watered with pivotal sprays are extremely wasteful of water. Not only is dehydration most intense in summer temperatures, but it is an extravagant way of applying water to crops. Drip irrigation is much more economical, but it must work efficiently and be applied daily if it is to reach the roots of plants sufficiently.

Drip irrigation, applied through small nozzles, also requires a water supply free of matter such as weeds and seeds. It must run freely if it is to water each plant adequately.

Furrow irrigation is nowadays decried as wasteful, yet it is a very efficient way in which to apply water to vegetable and fruit crops. The furrows are not deep, water is constrained from spreading to unproductive soil nearby, and water application is controlled easily.

Vegetables and fruit need to be consumed as early as possible from the time of harvesting as each minute after harvesting reduces the natural sugar content of the fruit and diminishes the nutritional yield.

Soft fruit such as raspberries and strawberries and freshly picked vegetables will use much more water if they are transported long distances, stored, washed, processed into precooked product and packed for sale in shops and supermarkets.

Fruit trees are grown under rainfall alone in temperate climates. In Australia, commercial fruit such as vines and peaches is irrigated as fruit size and sugar content must be maintained if the fruit is to be nutritious. In the growing period, which occurs in early summer, Southern Australia is mainly rain free and this lack of water will inhibit both yield and quality if trees are left to fend for themselves.

The major call on water in food made from fruit and vegetables is not so much from the growing period (this is a necessity), but during subsequent processing, packaging and storage. The longer fruit and vegetables are subject to these operations, the more the wastage from their natural water and sugars. All fruit and vegetables are best if lightly cooked and maintain their nutritional value if this takes place as soon as possible after harvesting.

5 Conclusion

In assessing the water content of food, we must distinguish between the natural requirements of the crop and the manner in which this food comes to our tables. Industrial processes that transform the natural product into a construct that looks much like the original but is treated with preservatives, chemically based aromas and various other additives such as extra sugar and salt require large amounts of water. Stretching meat into product to make more profit requires extra water.

An example of excessive and wasteful water is apparent in the German export of packets of iced tea to Italy. The ingredients are water and tea. Sugar and lemon flavouring are added for flavour. Italy has its own relatively ample water, buys tea from the same sources as Germany and grows fresh lemons that are exported to Germany, and sugar is an individual taste and better applied in the glass rather than the pack.

Almost every variety of food on sale in large commercial supermarkets requires large quantities of water to get it there. This use of water in wasteful does nothing to add to the natural qualities of the food and increases the price we pay for it.

The growing of food is always a risky business. Drought, hail, early frosts, searing winds and early snow can decimate a crop overnight. Irrigation can help with dry conditions, but we are at the limit of water available for irrigation. In any case, rivers and dams and other types of water storage are equally vulnerable to drought just when needed most.

Rainfall dryland farming sustainable in semi-arid regions has been an orphan in the development industry, yet it is in the semi-arid zone that the potential for increased food is most practical and achievable.

Unnecessary irrigation is wasteful, but the modern supermarket culture has introduced a need for water in food that goes far beyond the natural needs of the plant and animal.

While it is difficult to precisely measure the amount of water used by plants at any given time, it is easy to measure the extravagant use of water involved in industrial processes and food chains that distance farmed food from the consumer.

We need food to nourish ourselves and maintain good health. If we compromise these two values, then we diminish our own human condition and wastefully reduce our most needed resource that of water. As the water resource diminishes, the aim of food security becomes more unobtainable.

Science and technology can produce "cheap food" in abundance, but in the process, water is wasted, nutrition is compromised, and health is threatened.

When we talk about achieving food security, we need to understand what the consequences will be if we achieve this through large agro-industries producing industrial quantities of low-grade food. The alternative is to protect and sustain food that is grown in concert with nature and that reaches our tables while it is still fresh and highly nutritious.

Consumers in most of Asia, the Arab countries, Turkey and Southern Europe manage very well on fresh fruit, vegetables and meat sold direct to the consumer and cooked that day. The food culture in these regions depends on fresh ingredients marketed with little intervention to diminish their food value.

Food security is elusive, but it is better to aim for security in nutrition rather than agribusiness profit.

We in the Anglo-Saxon west have become seduced by the marketing of industrial food. Apart from demands made on the water resource, our food security is now threatened as adulteration, due to handling and storage and additions of unnecessary substances, continues to diminish nutritional quality. Some wealthy countries are now discovering nutritional deficits in children and old people due to this diet.

Agribusiness is designed to make profits. It is supported by technology that transforms produce into food components that can be manipulated for further profit. Transformed food has come to dominate our daily diet, not only in the West but increasingly and invasively in the rest of the world. It is a system that is water hungry, costly to the farmer, costly in terms of health to we the consumers and costly to our environment.

It may not be such a good idea to allow profit and technology to shape the food in which we should be secure. Increasingly, it is not the farming of food that threatens water supplies, but the processing of food that, it may be argued, makes more demands on water than the farming.

Farmers, farming medium size and small holdings and domestic plots using water efficiently to grow real food, may well prove to be our means of achieving food security after all.

Further Reading and Sources of Data

Blythman, J. (1988). *Bad food Britain.* Fourth Estate.

Chatterton, L. (2010). *From the ground up—Home cooking without fear.* Italy: Pulcini Press.

Chatterton, L. (2012). *Interventions and inventions in food production over the last century—The consequences of modern agriculture on the food we eat.* Paper Presented to the International Conference on Food Security in Dry Lands, Qatar.

Chatterton, L., & Chatterton, B. (1996). *Sustainable dryland farming, combining farmer innovation and medic pasture in a mediterranean climate* (paperback edition 2003, e book edition 2011). Cambridge: Cambridge University Press.

Elton, S. (2010). *Locavore.* Canada: Harper Collins.

Lawrence, F. (2004). *Not on the label.* Baltimore: Penguin Books.

Patel, R. (2007). *Stuffed and starved.* London, UK: Portobello Books.

Pollan, M., & Lane, A. (2008). *In defence of food.* London, UK: Penguin Books.

Shiva, V. (1991–2002). *The violence of the green revolution* (5 ed.). London: Zed Books.

Symonds, M. (1982 and 2010). *One continuous picnic—A history of eating in Australia.* London, UK: Penguin Books.

Water Sustainability and 0 km: Slow Food

Carlo Petrini

1 From Cars to Food Shopping: The Evolution of a Slogan

The idea of "zero km", originally regarded the sale of cars (and still does today), referring to cars that for some reason is registered and number-plated but had never been used, with some disadvantages (less colour and optional choices available) but many advantages (immediate delivery and much lower prices than listed prices) for buyers. An alternate means to the traditional one of buying a car, but with substantial savings.

The idea of applying this concept to food came about more or less ten years ago and can be attributed to Coldiretti when it began to become clear that a lifestyle geared towards sustainability could not leave a predominant consumption of local and (consequently) seasonal food out of the equation.

Food which, like the above-mentioned cars, does not travel kilometres before reaching the customer and, therefore, does not accumulate the different operator costs that occur along the supply chain, becoming a valid alternative to "conventional" supermarket shopping. However, for reasons that we shall see, the appealing and valid element in this choice does not lie (only) in a product's cost-effectiveness but in its higher degree of sustainability. Obviously, like all slogans, the idea of 0 km can appear to be a little approximate, schematic and not very realistic.

For example, if literally interpreted, it would appear to be unachievable in many cases—in large cities where those who have tried to organise markets selling local food products have found themselves forced to extend the 0 km up to at least 40 km to be able to realistically rely on nearby food sources, or in areas devastated by chemical or radioactive pollution, where it would be better for local people to avoid buying local food products until the area has been cleaned up.

C. Petrini (✉)
Torino, Italy
e-mail: c.petrini@slowfood.it

© Springer International Publishing Switzerland 2015

M. Antonelli and F. Greco (eds.), *The Water We Eat*,
Springer Water, DOI 10.1007/978-3-319-16393-2_6

At other times, however, where there is a quality local food production, the intricate web of regional regulations and services often results in a food product, that is produced and eventually sold within a radius of a few hundred metres, being transported hundreds of kilometres—production site to the processing plant and back to local selling outlet. An example of this can be found in meat products where the livestock farmers must often send their stock to abattoirs far from their farms before being able to actually sell the products to their local customers. Or processed products, such as cured meats or baked goods, that are sold and bought according to local consumer demand, but where their primary materials, animal feed or other ingredients, arrive from far away, weakening the importance both of the regional value and environmental protection. Where do the pigs for Italian ham come from? How can the organic poultry farmers in the south of Italy access certificated "Genetically Modified Organisms (GMO)-free" animal feed if they do not buy from the large companies in Lombardy?

Therefore, 0 km should be seen more as a *modus operandi* where the many variables are concerned, a range of nuances and possibilities, rather than as a precise and unquestionable description. Yet, whilst aware of the need to distinguish and specify, talking of 0-km food, or local consumption, is still a good way for us to understand or establish an objective. The success of an idea is also paid for in its simplicity and the need to remember, repeatedly and patiently, the complexity hidden behind a successful slogan.

2 The Idea of Sustainability Behind 0 km

From these first considerations, there clearly emerges a further "mitigating" element in the idea of 0 km. In fact, the very rough definition could be to refer to this idea as a particular type of commercial transaction, the purchasing (and consuming) of local food. However, from what has been said up to now, it can be clearly surmised instead that the idea of 0 km does not only concern the final step in the food supply chain, that is, the sale in a store or local market.

Zero km should not only be the final choice of those who buy food, but it should also, and firstly, be the choice of those who produce it. This should inevitably reflect on the choices made by those who manage the region and, therefore, by policy-makers at the local level and above.

In fact, the idea of 0 km involves aspects that concern the overall quality of food and its entire production process, beginning from the type of agriculture which this market, or that oriented towards sustainability, has as a target.

Above all, the idea of sustainability brings to mind the concept of "duration", forcing us to ask how and for how long can a process or behaviour last without creating any damage—on the contrary, possibly creating benefits—economically, socially and environmentally. Each of the latter can be further broken down—economic sustainability must yield to those who produce and those who buy, and it cannot be taken for granted that the two needs are necessarily opposite.

Those who produce must take into account the costs and see a return, but they must also be repaid for the services that they render to the environment or health or culture. This, of course, cannot occur only through pricing mechanisms, and it is here that the choices of the policy-makers come into play by supporting certain types of production. Those who buy must be able to have daily access to quality foods, but also have all the necessary information available to make well-informed choices and understand that buying food basically involves an investment in many different spheres. For example, by buying low-quality food products at low prices, we could find ourselves, in a not-so-distant future, forced to spend more on our individual health and, by means of taxes, on environmental health.

The problem of sustainability is the one we feel more "confident" about but we do not plan to spend more time on it here. We would like only to specify that the idea of environmentally friendly food must, once again, take into account different factors. In fact, it is not enough that food in itself is chemical-free, but its production must also contribute to safeguarding the natural resources and common assets which compete for its existence—water, air, soil fertility.

Finally, the social sustainability aspect regards compliance with the rules, respecting the rights and quality of life of all those involved in the production process. However, it also involves respecting the communities to whose culture a certain production is a part of, and which thanks to them has been preserved, evolving over time.

3 The Thinking on Waste Behind Water Sustainability

At this point, we can add a new element—water sustainability. It could have been included in the environmental sustainability previously mentioned, but for practical purposes it is better to deal with it separately and highlight it as an issue in itself. This is because if we begin asking ourselves about water, the issue is not only about renewable resources and their renewability cycles but it also involves our productive, distributive, commercial and consumption behaviours, and therefore, invariably, the issue of waste when these behaviours conform, without raising any intellectual resistance, to the dominant market rules.

On the issue of waste, a theme increasingly the focus of public attention, it is worth considering some further points for reflection.

Indeed, it is necessary to point out that where the so-called rationale of food production and distribution is concerned, waste is not an incidental element. Rather, it is a price to be paid. The faster production times, the quantities that must be produced to be able to distribute rapidly and widely, the marketing demands that call for bigger stores and, therefore, more and more goods, the various possibilities to increasingly lower production costs—all these factors involve, among their necessary evils, a certain amount of product waste and, consequently, resource waste. However, the term "waste" can be justified on at least two levels.

The first is its immediate and direct meaning—that is, from the moment that a consistent part of the food product (calculated at about 30 % considering all the steps, from production to consumption) does not fulfil its role (to feed someone) and is destined, depending on the situation, to become garbage or landfill, we may speak about waste in the "traditional" sense, or of something produced (and distributed, and bought) and then not used and thrown away.

The second concerns the alternative options in the chosen production process—if to obtain the same result I could have used less resources; if I choose to use more resources, it is then my choice regarding the waste produced, independently of the fact that the result (in this case to feed oneself) has been achieved. If the steak I choose to eat comes from an industrial livestock farmer and, therefore, has entailed a use of about 15,000 L of water, and if I had had an alternative, that is, to buy a steak originating from a sustainable livestock farmer who through his production and supply chain not only uses less water but actually works to protect water reserves, then I have made a "wasteful" choice even if I did not throw the meat away, but ate it.

It is not a coincidence that the other expression frequently used when speaking about water resources is water saving.

4 All the 0 km Steps

Therefore, in the light of the idea of water sustainability, let us try to retrace the steps in the selling/buying of a 0-km product, considering all the phases and beginning, of course, with the production in order to take into account all the aspects involved in safeguarding the resources which this type of transaction involves.

5 Traditional Seeds

The use of traditional seed varieties is among the main issues paradoxically absent from the Italian debate arena, even among the public more aware of food and environmental questions—somehow like water sustainability. Italy has made quality food an economic but also cultural and identity trademark. In fact, the peasant roots of Italians do not go so very far back in time. And to this, we can also add the growing awareness towards agronomic choices regarding sustainable food production. However, despite this promising scenario of increased competence and attention, the world of politics, research and, more generally, the civil seed society are hardly involved at all—it is not spoken about, there is very little literature and the existing regulations are inadequate. It seems we have forgotten that whatever the food production, including animal protein products (meat, milk, dairy products or fish), it all begins with a seed.

Instead, the initial difference between industrialised and sustainable agriculture actually lies in the seeds used. If the former uses so-called commercial seeds, which are aimed at improving a certain performance (usually quantity), the latter tends to use traditional seeds, the so-called landraces, which over the centuries and millennia have adapted to a certain area and respond efficiently to its pedo-climatic conditions.

In other words, commercial seeds are designed to be used in any area, as long as they are guaranteed the required external inputs, in terms of minerals and water and pest prevention or controls. Traditional seeds, instead, render the best results in geographically well-delineated zones, but in return they are more autonomous, in that they have adapted to a certain climate, and they have developed, thanks to continual selection by farmers, the best traits to yield a crop in that particular hot or cold climate, with scarce or plentiful water, in conditions that allow certain pests to proliferate in that poor, sandy or rocky soil. Thanks to this adaptation, which is nothing but evolution, traditional seeds require less care and less external inputs, and among these, of course, is irrigation. Therefore, choosing food from sustainable agriculture also involves being informed about the first step in the production chain —what seeds have been used? And the question is relevant, not only for buying fruit or vegetables, but also for animal origin products, as livestock farmers require fodder.

6 Agronomic Practices

Sustainable agriculture follows agronomic practices that, among other positive effects, reduce the need for irrigation. This happens, for example, with intercropping, with the presence, in the same plot of land, of crops that have complementary needs and roles, such as the input or fixation of soil nutrients, or which render reciprocal services, such as taller plants sheltering smaller plants intolerant to too much light or heat. Moreover, farmers who practise re-grassing (or who limit weed control) leave the wild plants to grow, for example, in kitchen gardens or vineyards, as they contribute to protecting the ground surface so that its natural moisture is conserved and its capacity to absorb rainwater is increased. All this reduces a cultivated plot of land's water requirement and improves the overall soil condition which will, consequently, be less prone to erosion, and more permeable during rainfall.

The harvesting phase also contributes to avoiding energy and water waste. Those who look to local markets or sell directly from the farm do not need to harvest and stock the product too early, thus saving in energy as well as water if we consider that storage facilities usually need cleaning, maintenance and air conditioning and air moisture controls. As far as livestock farming is concerned, where water requirements are involved, the industrial cultivation of animal feed (especially maize and soya) and the need for large intensive livestock farmers to sanitise stables and other facilities and machinery must also be taken into account. Instead, livestock farmers intent on sustainability have much fewer needs and even if they do not practise wild pasture grazing, they use fodder originating from sustainable agriculture and traditional seeds.

7 Packaging

The packaging of any food (and non-food) product is always very demanding in terms of water use. Whatever the type of polymer (from cellophane to polystyrene to PVC), paper, or any other solution, water consumption is an inevitable part of its production. To make the picture even worse, there is then the fact that a lot of packaging makes use of more than one system that many "plastics" are made up of different types of polymers, that the availability of the packaging for the industrial food products which are found in Italy and other countries almost always combines different material qualities. All of these factors result in different types of packaging. The main reason for this packaging system is the fact that the products must be transported, and therefore, safeguarded from any damaging effects. The commercial reason involving the wrapping of a product for an individual consumption, which can be found in a packaged product destined for the end-user, which, before, are part of a batch packaged for delivery to retailers, which, in turn, originate from larger batches of goods packaged for wholesaler delivery. The most classic example is that of cheese slices—each slice, for a single use is wrapped in cellophane; then, a pack of 10 or 20 slices are wrapped in plastic of a different chemical composition; then, packages of 10 or 20 of these are packed in boxes which are purchased by retailers; these boxes, 10 at a time, are placed in larger cardboard boxes; these are placed on a pallet, which for stability and safety, and are also wrapped in cellophane. If we compare this with a local farmer's cheese which is produced and matured in a natural environment and then transported to a market and sold directly to the end-consumer, you immediately realise that the latter is decidedly less demanding in terms of water use (and environmentally). The ratio between the product weight and the packaging weight, also considering it in simple weight terms and not "environmental weight", is diametrically opposite. But it does finish here. Because all packaging, sooner or later, ends up as rubbish, that may be sorted, and then recycled, or not. Certain waste disposal or treatment still requires water, even if waste decantation systems have been introduced which precede recyclable material recovery. However, once the primary material has been recovered, it is estimated that to produce 1 kg of recycled paper, 2 L of water is required, compared to the 100 needed to produce a kilo of new paper. Of course, in the first case the needs are very much reduced, but why not also eliminate those needs?

8 Shopping Places

How much water is used in building a supermarket? How much is needed in its daily operation? It is not an easy calculation to make, especially taking into account that also the word supermarket is becoming obsolete, rapidly being supplanted by more showy relatives such as shopping malls, hypermarkets and the like. Increasingly larger surface areas of land are being invaded by cement and warehouses and the places to shop are becoming poles to attract purchasers and suppliers who, in their turn, will need water to move and will consume water through their behaviour. And those places will have to be heated or cooled, cleaned, humidified… and every one of these actions carries a price in terms of water. The 0-km selling and purchasing places are the farms or farmer markets. Of course, you must move to reach them but, at least, the products have not had to be moved (it is estimated that the food we bring to our table daily has travelled about 1400 km!), and, anyway, the structures are already there, as in the case of farms, or they are less demanding structures, that do not ultimately take water from the region. If then, as well, you are lucky enough to be able to walk to do your shopping, even more water is reduced from our daily needs.

9 Waste

Buying local food also means developing an awareness of product value. This is not identified in or reflects the simple fact of its retail cost, as the consumer, like the producer, learns to see, also from a perspective of time, all the connections and benefits not only in terms of nutrition and health, but also in environmental, social and economic terms.

It is not only how you buy that changes but basically how you eat. For this reason, there is a tendency to increase the number of times you shop, reducing the number of purchases for each shopping trip. You plan more thoroughly based on your household food needs and you evaluate more carefully the possibility of wasting food. At the farm or market, there is no "below cost" policy or promotions, which tend to give the consumer the impression that the food they buy, and will possibly waste, has no real cost and, therefore, throwing it away is not a real waste.

This is why this type of buying is basically immune to wastage. This does not mean, however, that the 0-km products do not require water. Water is consumed, even if, as we have seen, to a lesser extent compared to industrial food products. However, basically, all the water required in their production will achieve its purpose, that is, to feed someone, and not end up in the garbage nor, ultimately, as landfill.

10 "We Support" Sustainability

No food production or behaviour exists that does not require water. But it is possible to decide how much water is used in our food and how much becomes waste through the choices made by each of the different food supply chain players who bring food to our tables. We are used to, thanks to the cultural conditioning the capitalist system and free market has exercised on each of us, considering and evaluating our behaviour and each choice we make, and even our every relation in economic terms. It is not always evident that we take this into consideration, often we refer to the need to "not waste time"; however, the evaluation we are making is an economic one. Beyond the evaluations that can be given to this type of competence, there remains the fact that we have become very skilled, experts in this type of evaluation. The new ability required at this moment in history and environmental scenario is to consider our actions, our choices and our daily behaviour in terms of water. How much water does it cost to go by train instead of by car? How much water does it cost, equal to the cost in money, for a lunch based on meat rather than one based on vegetables? How much water does it cost for an industrial snack rather than a cake baked at home? If we decide to adopt this reasoning, we will quickly realise that the different aspects of sustainability, which we raised at the beginning of these reflections, are linked. Obviously, this type of evaluation needs to be researched, so that the average citizen can have access to the information required to make a choice. And it is not random that we have decided to use here the term citizen rather than consumer. In fact, the term consumer encompasses a passive idea of eating behaviour (and not only) and especially excludes any possibility that those who make a choice in their buying can influence above all the production and, therefore, take a position, not as a consumer but as a co-producer. If we consider the need to acquire an awareness, still very distant, of water resources, it is clearly necessary to begin as soon as possible a thorough training and information dissemination at all levels—from political to research, and from production to purchasing.

The idea could be alarming, but we will easily realise that it is not such an unachievable goal if we think about how much and how deeply, in the last years, people have begun thinking about the idea of sustainability. Because if it is true that the issue of sustainability refers to many different spheres (economic, social, environmental), it is also true that only by achieving sustainability in each of those spheres, will we be able to guarantee a real sustainability regarding a product or behaviour. If, indeed, a production reveals itself to be environmentally sustainable, but economically disastrous for the producer, it would obviously be scrapped, with the result of gainsaying itself, that is, "not lasting". Or, if environmental and economic sustainability were to uphold social injustices or denial of rights, in the same way, that production or behaviour would be, of course, unsustainable.

For this connection that underlies all the different types of sustainability, a successful slogan can be found in Slow Food's "good, clean and fair", highlighting a food product's overall quality. It is becoming increasingly clear, at all levels, that what Fritjof Capra called, in the 1970s, "the hidden connections" must become a focal point. And among these connections, there is also the one that concerns water, its availability, its renewable cycles and, therefore, what can be consumed, the balances to be safeguarded and the availability to be guaranteed, locally and globally.

11 Let Us Tell All

For all these reasons, it has become fundamental, as has been mentioned many times in these pages, but is worth highlighting again, that those who make the choices in buying must be able to access the information regarding the production process of the good they wish to buy. And for this, producers must be equipped with suitable means to be able estimate their products' environmental impact and water costs. This means that research must dedicate more energy and time to these studies and politics must dedicate more funds to this research. The water footprint accounting of a food product must be one of the criteria for evaluating its overall quality.

Virtual Water in Diet, Shopping and Food Waste

Andrea Segrè, Luca Falasconi and Cecilia Bellettato

1 Food Consumption

Food is the central pillar for the life of humankind; food is also an important element of our history and culture. In addition, food is an essential part of the environment and the places it originates from. Ludwig Andreas Feuerbach, a German philosopher, in one of his famous aphorism, expressed the idea that "We are what we eat". If this concept is to be still considered valid, this means that the path we are undertaking is rather dramatic. In order to demonstrate that, it will be sufficient to articulate just a couple of considerations. The first is that, in Europe, 43 % of food waste is domestic waste. The second is that food is gradually becoming simply a commodity, that is, a good which must be traded at the lowest price, losing not only its economic value but also its nutritional, cultural, social and historical values.

This loss in importance, on the contrary, is in direct conflict with its global and steadily growing demand. This aspect is a major worry for analysts, in view of the projections forecasting 9 billion inhabitants who will populate our planet by 2025. Demographically, the world's barycentre will progressively shift towards developing countries and towards cities. By 2050, the African continent will have a population of 2 billion, doubling its present numbers. By then, it is also foreseen that two-thirds of the population will live in urban areas. This latter aspect, linked to the variations in the demographic structure of rural areas, will result in a 30 % decline in populations working in agriculture (United Nations 2008). If these predictions will actually reveal true, the demand for food from now to 2050 will inevitably tend to polarize into two opposite directions. On the one hand, we will

A. Segrè · L. Falasconi · C. Bellettato (✉)
Alma Mater Studiorum, University of Bologna, Bologna, Italy
e-mail: cecilia.bellettato@unibo.it

© Springer International Publishing Switzerland 2015

M. Antonelli and F. Greco (eds.), *The Water We Eat*,
Springer Water, DOI 10.1007/978-3-319-16393-2_7

have food whose sole purpose will be to satisfy the nutritive needs of low-income populations and therefore based on poor quality products (from a mere economic point of view); on the other hand, we will have a nutrition model that will be enriched by many additional services, aimed at meeting the growing needs for an improved accessibility of food products for those living in metropolitan areas.

These trends will lead to an increase in commodity goods consumption. The FAO estimates that by 2050 we will be consuming an extra one billion tonnes of cereals, a little less than 200 million tonnes of meat, about 660 million tonnes of roots and tubers, 172 million tonnes of soy, 429 million tonnes of fruit and 365 million tonnes of vegetables. This will imply an increase in consumption of 60 % compared to 2005–2007 (Cerosimo 2011). In addition to this quantitative increase, an increase in quality will also occur. In many countries, particularly in the so-called developing ones, there will be a noticeable increase in the consumption of products from animal origin and also of sugar and vegetable fats. Globally, however, the main energetic source will remain that supplied by cereals, from a minimum input of 15–30 % of the total caloric intake in those countries where there is a high consumption of roots and tubers and up to a maximum of 70–80 % in those contexts where there is a high consumption of millet and sorghum (Africa) or rice (Asia) (ibidem).

Developed countries will contribute to maintaining their high levels of cereal consumption, not so much due to their direct use but for their use in animal feed. According to FAO estimates, by 2050, the amount of cereals for direct human consumption will be surpassed by that for other uses and, specifically, for the quantity used in animal feed. Concerning the consumption of products from animal origin, as already mentioned, an increase in consumption is also, and especially, forecasted in those countries where this appears to be still quite contained, including India (which in 2005–2007 was below 10 kg per capita a year). However, all the above must be evaluated in the light of the fact that even if income levels are gradually increasing in these countries, they are still notably lower than in Western countries. In other emerging countries, such as China and Brazil, in the coming decades a slowing down is forecasted in the increase in animal origin product consumption. This is due to the fact that they will approach a situation of maturity of their consumption patterns. [1]Given that this threshold has been exceeded in developed countries, this consumption will also decrease further for reasons of health and environmental sustainability. Italy, even if maintaining the particularity of its traditions of a unique food culture, will shift in the above-mentioned direction. Evidence of this fact can be found in the data regarding the average spending for Italian families, which, from a quantitative perspective, is in line with families of other European countries (EU 15). What is different, however, is the composition of the average shopping basket (Cerosimo 2011). Here, we need to underline how, in

[1]The characteristics of food consumption in a mature and sated society are the saturation of energetic foods, the saturation regarding food shopping and a certain tendency to a similar diet from a nutritional aspect, but with differences in the qualitative nature and in consumption behaviour.

the last decade, the spending of Italian families on food consumption has undergone a noticeable decline. Comparing the data on consumption from the beginning of the new millennium with those of today, the percentage of spending for food has greatly decreased and has returned to the level of the early 1980s. However, concerning this trend we must also consider the increase in spending for restaurant and catering services. In 1992, average household spending on food was 18.3 % of total spending, while for restaurants/catering it was approximately 6.5 %. In 2011, this percentage increased to 14.52 and 7.9 %,[2] respectively. In particular, the categories that recorded a drop in consumption were meat,[3] oils and fats (OF)[4] and fruit.[5]

There are basically three reasons underlying these changes in preference and eating habits (Peta 2008):

1. socio-demographic changes (ageing of society, increase in single-person households, multi-ethnicity) associated with lifestyle changes;
2. more attention to diet due to pathologies linked to types of consumption (obesity, food intolerances and allergies);
3. inequalities in incomes and purchasing power of Italian families (Italy is one of the European countries with the highest Gini Index[6]). The percentage of income spent on food consumption is higher among those from lower income groups, and vice versa.

Since, in Italy, the gap between the wealthy and the poor is increasingly widening, it is foreseen that in the near future trends in food consumption will also diverge. On the one hand, there will be a demand for high-quality products and services, and, on the other hand, the purchasing of low-price food products.

2 The Agri-Nutritional Models and Their Evolution Due to Changes in Consumption Behaviours

The constant and steady evolution in lifestyles and food consumption can also be confirmed in the study of the agri-nutritional models proposed by Louis Malassis. The agri-nutritional model (MAN), developed in France by Malassis and Padilla

[2]ISTAT (2012), *Rapporto annuale 2012. La situazione del paese*, Rome; Id. (2011), *Annuario statistico italiano 2011*, Rome.

[3]This product in the family shopping for food products has decreased from 25.4 % in 1992 to 22.1 % in 2010.

[4]This product in the family shopping for food products has decreased from 4.9 % in 1992 to 4.1 % in 2010.

[5]This product in the family shopping for food products has decreased from 7.3 % in 1992 to 6.8 % in 2010.

[6]The Gini Index measures the inequalities in wealth or income distribution of a country. It is a number that varies from 0 to 1, where 0 corresponds to a fair and equal distribution, while higher values correspond to higher levels of inequality.

Table 1 Food categories

Abbreviation of basic product groups	
Cereals and starch products	CSP
Sugar and sweeteners	SS
Fruit and vegetables	FV
Legumes	L
Meat and eggs	ME
Fish and seafood	FSF
Milk and dairy products	MDP
Oils and fats	OF
Alcoholic beverages, spices and stimulants	A

Source The authors

(1986),[7] is the nutritional system of foodstuff availability by volume of a given country. At the end of the 1980s, the two French agri-food economists analysed the dietary habits of consumers in European countries and the main non-European countries with the purpose of classifying them based on their agri-nutritional model.

In order to define them, it was first necessary to categorize basic foodstuff availability into homogenous product groups, as seen in Table 1.

Based on these groups, the available agri-food resources in each country are calculated using food balances, which establish, for each product category, the availability in kilograms, final calories and nutritional elements available in each country in a given year. As a consequence, it is possible to obtain the daily availability in grams and final calories per inhabitant. The agri-nutritional models of specific countries are thus defined by agri-nutritional availability indices expressed in final calories per product group. This results in highlighting the relative importance of each group in the energetic structure for each country covered by the study. However, in determining the agri-nutritional model, only those product categories are considered with more than 10 % of the average availability per inhabitant per day of the total caloric intake. This method therefore gives priority to the energetic features of a food product, rather than to the quantitative ones.

Following this approach, four models can be defined, establishing their nutritional characteristics using a mean value of the data of countries belonging to the same group:

- the traditional agricultural model characterized by only two food categories which contribute more than 10 % kcal to the diet, i.e. cereals and starch products (CSP), OF;
- the Mediterranean model characterized by the presence of more than 10 % kcal deriving from CSP, OF and meat and eggs (ME);

[7]Malassis and Padilla (1986), *Economie agro-alimentaire. L'économie mondiale*, v. III. Cujas, Paris.

Table 2 Country categories based on their agri-nutritional model—1975–1977 and 2007

Typology	Food characteristics	1975–77	2007
Anglo-Saxon	CSP–SS–MDP–ME–OF	Denmark, Finland, Ireland, Holland, UK, Sweden	Estonia, France, Germany, Ireland, Holland, UK, Sweden, Argentina
European	CSP–SS–ME–OF	Benelux Federation, France, Germany	Austria, Cyprus, Denmark, Malta, Lithuania, Poland, Czech Republic, Hungary, Brazil, Chile, Colombia, Paraguay, Suriname, Uruguay, Venezuela
Mediterranean	CSP–ME–OF	Italy, Portugal, Spain	Italy, Portugal, Slovenia, Spain, Ecuador
Agricultural/cereal	CSP–OF	Turkey	Albania, Turkey, Peru

Source The authors

- the European model which adds sugar and sweetener products (SS) to the above-mentioned three categories;
- the Anglo-Saxon model, the most complete, which contributes more than 10 % kcal of the CSP, OF, ME, SS and milk/dairy product (MDP) categories.

The four agri-nutritional models have a more or less similar overall caloric input[8]; the difference results from the variety of eaten food. In fact, as can be noted, the Anglo-Saxon diet is the most complete, and it is the only one which includes more than 10 % of MDPs, while, for example, in the traditional agricultural model, few calories from animal origin are consumed. In the Anglo-Saxon agri-nutritional model, a good 43 % of calories are of animal origin, whereas in the agricultural model, this category is less than 10 %. In all models, except for the Anglo-Saxon one, we find a predominance of carbohydrates and lipids. It is quite clear that in all the models we find fruit and vegetables, legumes, fish and seafood, alcoholic beverages, spices and stimulants, but their total caloric input is less than 10 %; therefore, they are not considered as food categories characteristic of the agri-nutritional model (Malassis 1995) (Table 2).

From 1975–1977, with a higher kcal consumption per capita, due to an increase in incomes, there followed more diversification and a balancing in diets (less kcal originating from cereal/starch products and an increase in animal origin kcal consumption) without, however, a great revolution in eating habits occurring (Segrè and Falasconi 2012).

[8]An average calorie intake of approx. 3500 kcal, the only exception being Albania where daily consumption in 2007 was approx. 2900 kcal.

3 Water Consumption in the Different Agri-Nutritional Models

As far as the impact that each agri-nutritional model has on water consumption is concerned, three case studies for each agri-nutritional model type were carried out. The countries studied were as follows: Albania and Turkey for the traditional model; Italy, Spain and Portugal for the Mediterranean; Denmark, Austria and Cyprus for the European; and France, Germany and the UK for the Anglo-Saxon model.

Using the food categories composing each agri-nutritional model (CSP, OF, ME, SS, MDP), first of all, the water weight of each model for each country was calculated as a sample, using the global average of virtual water contained in the selected foods. Secondly, the water weight of each of the four models was estimated using a weighted mean of the results obtained. The units of measurement used as a reference were the cubic metres of virtual water per product tonne. The data relevant to the virtual water values were taken from the database created by Mekonnen and Hoekstra, of the Water Footprint Network (Mekonnen and Hoekstra 2011).

To provide an example, for the "cereal and starch product" category, the virtual water contained in wheat, rice, barley, maize, rye, oats and potatoes was considered. The products included in the food categories characteristic of the country's agri-nutritional model were calculated as an integral part of the diet. For these products, thanks to the FAO database, we found the kg per capita consumed in a year (2007) per country, which, multiplied by the cubic metres of virtual water required to produce a kilo of product, eventually provided us with the water footprint of the consumption of that food in that country. For Italy, for example, those foods included in the "sugar and sweetener" and "milk and dairy product" categories were not calculated as an integral part of the diet, as they are not characteristic of the Mediterranean agri-nutritional model to which Italy belongs. The water impacts of the products considered for each country were summed up providing the energetic impact per capita of the diet for each country. Then, the weighted mean was calculated for the countries belonging to the same agri-nutritional model, so as to have not only the weight of each single country but also of each different model (traditional agricultural, Mediterranean, European and Anglo-Saxon).

The results are the following: the traditional agricultural diet consumes 439 cubic metres of virtual water per inhabitant a year; the Mediterranean consumes 1715; the European 1934; and the Anglo-Saxon 2607. Therefore, there is a considerable difference between the traditional diet, in which agricultural products prevail (cereals, starch and oil products), and all the other diets. The Mediterranean, with the closest virtual water content, consumes almost four times more, the Anglo-Saxon almost seven times more. The huge difference is due to the absence of animal origin products, except for animal fats in the traditional agri-nutritional model. As previously mentioned, animal origin products have a very high water footprint because of the fodder consumed by the animals (Mekonnen and Hoekstra 2011). It is worth underlining how this difference has a totally different impact from an environmental point of view, based on how the animals are bred and raised (grazing where they feed on grass in open fields; or feeding in closed stalls). The difference between the Mediterranean agri-nutritional model and the European one is instead not so marked, as, in this case, only sugars and sweeteners, vegetable products, are added. Finally, the Anglo-Saxon diet consumes a much higher quantity of water, even compared to the European one, as milk and dairy products are also included and they have quite a high virtual water footprint.

These studies highlight how consumers can play an important role in making an effort to reduce the level of global water use. In changing their dietary habits, a challenge to be faced in order to achieve this goal, consumers would not only benefit the environment but also their own health. This makes it increasingly more convincing and obvious that a small commitment from each one of us can lead to remarkable results without any real great efforts being made and without having to change a great deal, and not even revolutionize, our eating habits and behaviour. What is really needed is a little bit more of attention, from each one of us. These small efforts, even if adopted by one single person, can lead to important results.

4 Waste Within Waste

After having analysed the different agri-nutritional models and having compared how food habits play an important role, not only for our health but also for the environment and the world's natural resources, we will now move into looking at what we will define as the waste within waste, namely the waste in resources occurring when food is wasted. If, in fact, any type of food consumption can be justified at the moment in which the produced food is utilized for its end use, and thus human consumption, it becomes quite unacceptable that natural resources, such as water, land, energy and work, are used to produce a food products that will not end up, for various reasons, in being consumed. An enormous quantity of still perfectly edible food is currently thrown away.

Before going any further, it is important to establish two concepts, namely food waste and, more specifically, the waste regarding agricultural products and what we consider the water waste linked to this phenomenon. Where food waste is concerned, no single definition exists. Instead, there are many different and more or less broad definitions which include the different phases along the food supply chain and the different typologies of food products. Therefore, it is worthwhile in this case to specify what exactly should be identified as waste. "Food waste" is the sum of all those perfectly usable products that, for different reasons, are no longer sellable and that, lacking in any possible alternative uses, are to be eliminated or disposed of. Food products that constitute waste lose their characteristic of being a "commodity", but not their main feature: the characteristic of being still "food". Therefore, they are unsold but not unsellable products (Segrè and Falasconi 2012).

Waste occurs at every point along the supply chain, from production to the end-consumer, passing through the steps of the first and second processing process, transport, wholesale and retail selling. Where Italy is concerned, the data gathered and re-elaborated [9] show that the agricultural sector is one of the areas that has the highest waste factor. The reasons why waste is created are various, ranging from commercial issues to market logics or linked to the producer's strategic choices. On an agricultural level, food waste can be divided into three macro categories:

1. waste linked to crops that are not harvested because they are aesthetically damaged by parasites, diseases or adverse weather conditions (hail, ice, heat waves, etc.);
2. waste linked to economic reasons. If, when harvested, the market prices are too low, the farmers can leave a part of their crops in the fields as their harvest and successive sale would not be enough to cover the production costs (in some cases, the market prices are not even enough to cover the costs of the harvesting work);
3. waste linked to the product's commercial defects. Goods that do not meet the minimum quality standards regarding shape, size, colour and ripening times are left in the fields.

There could also be a fourth cause linked to the fact that farmers, in order to avoid losses due to adverse weather conditions or parasite problems, or "gambling" at the beginning of the farming year on higher final market prices, try to increase their yields. However, in good years (where there are no weather, parasite or disease problems) there is a surplus yield which for the above-mentioned economic reasons is not harvested, remaining to rot in the fields. Usually, these unharvested crops are ploughed back into the soil. This highlights the fact that we are not looking at a complete loss as the organic matter is recycled into the land. However, surely this behaviour is, to all extents, a nutrient loss for humans. As previously defined, the product loses its "commodity" feature but not its "food" feature. Therefore, if it is ploughed back into the soil it literally becomes waste.

As far as the second definition is concerned, that is, the water waste linked to food waste (the same discourse is applicable to all the other resources used in the productive process—energy, work, fertilizers, seeds, pesticides, etc.), the reasoning becomes more complicated. In fact, we could consider that at the moment the resources used for the production of a good result in bringing that good to maturation in the field, they have, from a technical point of view, achieved their purpose and so have not been wasted. In following this reasoning, the product's final use is of no importance as it can be used for different purposes—food for humans or animals, the production of biofuels or compost or also ploughed back into the ground. However, we have affirmed that the end point of food is human consumption. Consequently, the moment food is marked for a purpose other than human consumption, all the resources that have been used to produce that good have not achieved their final goal (from an economic and social point of view) and, therefore, have been wasted. Thus, it is with this reasoning in mind that we intend to

[9]For further information, see Segrè and Falasconi (2011).

speak about waste within waste, that is, resources that were used to produce the goods that then reached the end-consumer's table. Waste and waste within waste, as we have just affirmed, are very widespread phenomena in Italy, but not only.

If we take Italy as a case study, we can see how, only in 2010, 15 million quintals of agricultural products remained unharvested in the fields—3.2 % of Italian production. These data, is given in (Table 3) which as a percentage could almost be considered physiological, if considered in its absolute value clearly shows how a little more than 12.85 million quintals of fruit and vegetables and 2 billion quintals of cereals were left in the fields: more specifically, 307,887 quintals of peaches (excluding nectarines), 1,560,992 quintals of oranges and 1,348,515 quintals of grapes were left to rot in the fields. This means that in Italy a relevant quantity of fruit and vegetables, slightly less than a tenth of what gets consumed, were wasted; that is, every ten kilos of fruit and vegetable consumed by an Italian has a tare weight of almost a kilo of what is left to rot. From an environmental point of view, this means having used resources, including water which is a fundamental, but at the same time scarce, good for human life, to produce waste or, even worse, garbage. It is clear that this is difficult to accept both environmentally and ethically, in a world where many people still not only suffer from hunger but often do not even have access to resources that we are widely wasting. These resources could have been used differently or better, simply conserved, in a sustainability perspective for future generation. The same sustainability perspective everyone talks about in the last few years.

Focusing on the water resources using the above-cited data, we can observe that in 2011 a little more than 1.2 billion cubic metres of virtual water were wasted, more or less the quantity of water in Lake Iseo in the north of Italy. Specifically, 13,851,139 cubic metres of water were used in the cultivation of peaches left to rot in the fields, 58,499,890 cubic metres for oranges and 89,803,337 cubic metres for grapes.

Fruit, with 69.5 % in losses, registers the highest waste in the sector, followed by cereals with 22 % and, finally, vegetables with 8.5 %.

Table 3 Italian agricultural production in 2010 and virtual water wasted

Prodotti	2010				
	Produzione Totale (q.li)	Produzione Raccolta (q.li)	Residuo in campo (q.li)	%	Totale acqua virtuale (m^3)
Totale Cereali	171,327,090	169,051,949	2,275,141	1.33	264,698,098
Totale Frutta	213,318,127	205,274,297	8,043,830	3.77	836,511,414
Totale Orticole	87,465,678	84,508,099	4,809,731	5.5	102,732,389
Totale Ortofrutta	**300,783,805**	**289,782,396**	**12,853,561**	**4.27**	939,243,802
Totale	**472,110,895**	**458,834,345**	**15,128,702**	**3.2**	1,203,941,900

Source The authors

Products	2010							
	Green	%	Blue	%	Grey	%	Total	%
Fruit	7,284,400,790	72.6	723,385,281	68.5	357,328,066	37.8	8,365,114,138	69.5
Vegetables	655,267,738	6.5	228,390,198	21.6	143,665,950	15.2	1,027,323,886	8.5
Fruit/Vegetables	7,939,668,528	79.1	951,775,479	90.1	500,994,017	53	9,392,438,024	78
Cereals	2,097,686,019	20.9	104,643,738	9.9	444,651,223	47	2,646,980,980	22
Total	**10,037,354,548**	**83.4**	**1,056,419,216**	**8.8**	**945,645,240**	**7.9**	**12,039,419,004**	

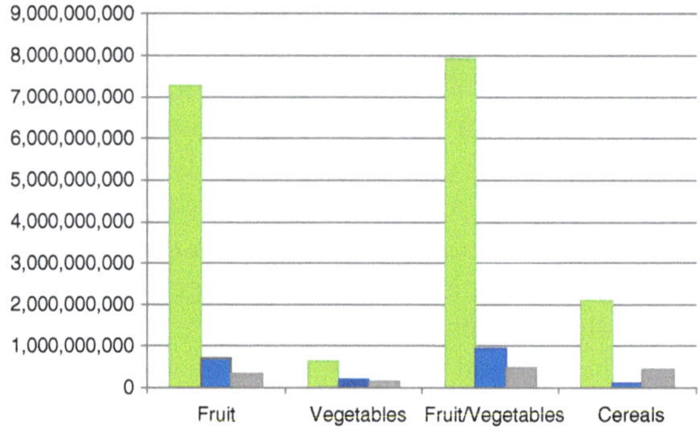

Fig. 1 The 3 virtual water elements linked to waste in the fields in 2010. *Source* The authors

However, besides the given total it is interesting to analyse the waste of the different typologies of water (blue, green and grey). In fact, one thing is to waste rainwater, which has a very-low-cost opportunity, as it is difficult to use for purposes other than agriculture; another thing is to throw away blue water which could be used in many alternative ways and which, sometimes, originates from non-renewable sources. Yet another thing is to pollute, more or less high volumes of water (grey virtual water) through fertilization or pesticide use.

Figure 1 shows how the fruit/vegetable sector has the greatest impact in terms of blue and grey water waste. The blue and grey water wasted in Italy in 2010 could have satisfied an estimated domestic consumption of almost three million people for the entire year, 5 % of the total Italian population (considering the current consumption which many studies show to be more than the actual needs).

These figures, already quite worrying, only refer to the first step in the food supply chain. If only the data banks allowed it, we could analyse the waste throughout all the steps that food takes from the field to our tables. These data would be quite alarming, as well as being unacceptable economically, environmentally and ethically. It is precisely for this reason that the question of food waste is vitally important and it is fundamental to raise awareness around this issue where even small reductions in waste can have very important consequences and result in notable benefits in many aspects. Reducing waste even more than changing the agri-nutritional model would not involve any kind of sacrifice; therefore, we feel that it is an "effort" that could be made and should be asked to all consumers. A commitment that would mean, for each consumer, being just a little more careful about how they eat. Eating is an action that, if we think carefully, allows us to survive (besides providing enjoyable moments of conviviality and personal fulfilment) and, therefore, "deserves" an "effort" of care by each and everyone of us.

References

Brunetti, A., Felice, E., & Vecchi, G. (2011). Reddito. In G. Vecchi (Ed.), *In ricchezza e in pvertà. Il benessere degli italiani dall' Unità a oggi* (pp. 209–234). Bologna, Mulino.

Cerosimo, D. (Ed.). (2011). *I consumi alimentari. Evoluzione strutturale, nuove tendenze, risposte alla crisi*, findings from the workshop held at Palazzo Rospilgioso 27 September 2011, Gruppo 2013. Quaderni, Edizioni Tellus, Rome.

Hoekstra, A. Y., & Cjapagain, A. K. (2008). *Globalization of water: Sharing the planet's freshwater resources*. Oxford: Blackwell Publishing.

Id. (2012). *Rapporto annuale 2012. La situazione del paese*. Rome.

Id. (2012). *Libro blu dell spreco in Italia: l'acqua*. Edizione Ambiente, Milan.

ISTAT. (2011). *Annuario statistico italiano 2011*. Rome.

Malassis, L., Lanini, L., & Ghersi, G. (a cura di). (1995). *Introduzione all'economia agroalimentare* (pp. 424). Bologna: Il mulino.

Malassis, L., & Padilla, M. (1986). *Traité d'économie agro-alimentaire. L'économie mondiale* (Vol. III). Cujas, Paris.

Mekonnen, M., & Hoekstra, A. Y. (2011). *National water footprint accounts: The green, blue and grey water footprint of production and consumption* (Vol. 2(50)). The Netherlands: UNESCO-IHE.

Peta, E. A. (2008). *Consumi agroalimemtari in Italia e nuove tecnologie*. Awareness-raising campaign, Ministry for Economic Development; http://www.dps.tesoro.it/documentazione/uval/Consumi%20e%20Innovazioni%20Marzo_rev.pdf.

Segrè, A., & Falasconi, L. (2011). *Libro nero dello spreco in Italia: il cibo*. Milan: Edizione Ambiente.

Segrè, A., & Falasconi, L. (2012). *Il libro blu dello spreco di acqua, edizione ambiente* (pp. 201). Milano.

United Nations Economic Commission for Europe. (2008). *Standards for perishable foodstuffs*. Paris: OECD.

Aware Eaters of Water: An Idea for Water Labelling

Francesca Greco and Marta Antonelli

Abstract Food represents 90 % of the water consumption of an individual and the agricultural sector uses on average 70 % of the freshwater withdrawn from surface and aquifers for irrigation purposes globally. From the perspective of a sustainable growth, oriented towards optimizing the use of green water and reducing that of blue water, i.e. irrigation waste and inefficiencies, it is essential to raise citizens' awareness and promote more sustainable consumption. For this purpose, this contribution will discuss the possibility of guiding the commercial choices we, the citizens, make by means of a method for labelling water sustainability. This hypothesis of labelling provides "qualitative" information on the typology and origin of the water used to produce any type of food we consume.

1 Labelling Environmental Sustainability

The preference of a consumer between two and three alternative products is usually given to the good whose value is perceived as higher compared to any other available choice. The environmental information, provided by the product's label, but not only, is surely one of the factors that influence the consumer's ability to make an aware decision, conditioning the final choice. According to the Italian Law Dlss 195/05, an "environmental information" is "any information available under

Both authors contributed equally to this work.

F. Greco (✉)
King's College London, London, UK
e-mail: francescagreco78@gmail.com

F. Greco
United Nations World Water Assessment Program (UN WWAP UNESCO), Perugia, Italy

M. Antonelli (✉)
IUAV University of Venice, Venice, Italy
e-mail: martaantonelli84@gmail.com

© Springer International Publishing Switzerland 2015

M. Antonelli and F. Greco (eds.), *The Water We Eat*,
Springer Water, DOI 10.1007/978-3-319-16393-2_8

written, visual, auditive, electronic form, or under any other form, regarding the state of the water, air, soil, territory, coastal and marine zones, including the biodiversity, the genetically modified organisms and any other factors, activities and measures destined at safeguarding the environment, including administrative measures and environmental agreements".

Environmental labelling is a part of the information category called *market-oriented* information.[1] It involves a "statement indicating the environmental aspects of a product or service" (Crivellaro et al. 2012, p. 178; Regulation ISO 14020) and is aimed at raising consumers' awareness of the actual value of that good, informing them about its distinguishing features which would otherwise be completely invisible to the eye (Wales et al. 2010). These "environmental" characteristics refer to what happens in the production chain, in the phases prior to their purchase (Boström and Klintman 2008). In other words, the environmental label must translate and summarize the environmental complexity into a brand or symbol which is simple and reliable. Certification standards have been recognized as a way of influencing not only private sector operators, but also as a means to respond to consumer demand (RPA and Cranfield University 2011), which would also result in reducing the time taken to seek information regarding the environmental impact of commodities (Boström and Klintman 2008; Teisl 2007).

The opportunities linked to adopting an environmental label for water resources has also been recognized at the European level. The July 2007 Communication "Addressing the challenge of water scarcity and droughts in the EU" confirmed that the adoption of *ecolabels* is one of the means to improve water resource management in the member states. Similarly, a report was drawn up by the European Commission in 2011, "Assessing the efficiency of the water footprinting approach and of the agricultural products and foodstuff labelling and certification schemes" (RPA and Cranfield University 2011), which encourages the use of the concepts of virtual water (defined as the volume of water required to produce a good) and water footprinting (defined as an indicator of the water consumption of a given subject or group of subjects in a given period of time) as tools to inform labelling and certification schemes to be placed on agri-food products in order to reorient water resource planning and management leading to more efficient use and savings. The report underlines the need to improve the consistency, clarity and transparency of the calculation methodologies and the use of indicators.

The reference to virtual water and water footprinting is also found in the following: "Plan for safeguarding European water resources" in which the Commission supports "the use of awareness-raising tools such as communication campaigns, and

[1]Concerning environmental information, it is important to distinguish between two categories. The first is defined as *compliance-oriented* and refers to all the information for administrative–bureaucratic purposes which are not aimed at influencing consumer choice in purchasing but are addressed to the control institutions or entities. The second category, instead, refers to *market-oriented* information, with the purpose of influencing the consumer to take, when purchasing, the most aware decision possible. Labels, just as information spread through advertising or found on product packaging, are also included in this category (Crivellaro et al. 2012).

certification and water footprinting schemes to provide incentives for water users to make sustainable choices" (European Commission 2011 p.19). It proposes to raise the awareness of all member state citizens to the problems inherent in water consumption through a system of labelling and certification to be carried out on a voluntary basis.

2 The Experience in Italy and the World

Internationally, many initiatives already exist aimed at monitoring and communicating a sustainable water management throughout the supply chain. Among its initiatives, the Water Footprint Network proposed the "corporate water footprint" drawing up a manual for private sector operators (Hoekstra et al. 2011). The objective is to establish universal standards to calculate the water footprint in all phases of the supply chain. There are three main key points in the manual:

- promoting a transparent calculation of water use in the different supply chain phases;
- considering the water footprint and its social, economic and environmental impacts;
- mitigating the impacts through methods, reducing water use and fostering an increased efficiency.

The Water Footprint Network initiatives have also been extended to actual training courses for company employees.

The first non-governmental organization to become involved in safeguarding and managing water resources (water stewardship) was the WWF International (WWF 2009). In particular, the WWF United Kingdom boasts an experience of many years alongside the Water Footprint Network (Chapagain and Orr 2008). Within this network, a working group, made up of WWF members but also including UNESCO_IHE, Suez and the "World Business Council for Sustainable Development" members, was able to involve and obtain the commitment of some private giants, such as Coca-Cola, SAB-Miller, USAID, TNC and Nestlè (http://www.nestle.com/csv/water/collectiveaction). Another important partnership is the "Alliance for Water Stewardship" (members include the WWF, UNEP and the CEO Water Mandate and Water Witness International). Among the many other experiences, it is worth mentioning the FAO initiative on food product labelling (for further information see http://bit.ly/YtYzWM) and WWF Finland which is very active nationally, also with education campaigns in schools (see http://bit.ly/12ludN6). Finally, the public–private initiative of the above-mentioned CEO Water Mandate should be underlined, being the United Nations coordinating group (UN Global Compact) which is part of the "Alliance for Water Stewardship", together with the Swedish government and some private groups. In 2007, these institutions joined together to launch a challenge to promote reducing the water footprint in the industrial sector.

The creation of the ecolabels in Italy is still underway. The consultancy group SPRIM and researchers from the Institute for Agrarian and Environmental

Chemistry of the University "Cattolica del Sacro Cuore" have designed a "multi-criteria" ecolabel to be used for consumer goods. This label indicates a good's environmental impact on the air, water and earth identifying 18 indicators grouped into three macro-areas and using the life cycle assessment methodology (life cycle assessment; http://bit.ly/X7PPsc). The comparative factor of the scale is the average daily environmental impact of the European citizen. However, if using an average of European consumption as a reference could be acceptable in preparing a label to be disseminated all over Europe, even more interesting would be to compare the environmental footprint of the different countries with a global average, as the Water Footprint Network proposes for the water footprint (Mekonnen and Hoeskstra 2011). This comparison with a global average would, in fact, result in highlighting, for example, how the European and North American water consumption model is well over the world average.[2] Therefore, how to take into account different levels of consumption when designing an ecolabel is not such an insignificant problem. Among the more interesting proposals, we can mention the Slow Food's "narrative label" presented during the 2012 *Salone del Gusto* in Turin. This label is actually a product identikit as it includes an exact description of the characteristics linked to the region and production period, the origin and diet of the livestock (in the case of animal products) and the farming methods of the product in question (http://www.fondazioneslowfood.it). Raising consumer awareness of the impact that our daily consumption choices have on the environment is also the objective pursued by WWF Italia which, in cooperation with the University of Tuscia, the University of Naples and the tomato company Mutti, has proposed on its Website "the virtual shopping trolley" (http://www.improntawwf.it/carrello), a calculator of the environmental impact allowing us to obtain, using an "environmental receipt", the price of the "bought" food, not in monetary terms but in the cubic metres of water consumed and the CO_2 emissions produced. The COOP supermarkets have also adopted a similar idea where its "shopping footprint" (http://www.coopambiente.it/) allows the user/consumer to calculate their CO_2 footprint on a low–medium–high impact scale.

Finally, besides a committed academic world which is increasingly focusing its attention on the issues of virtual water and water footprinting,[3] WWF Italia, Slow Food and different private players, such as Barilla, we would like to mention the work of the scientific dissemination carried out by the info-designer Angela Morelli, both in Italy and abroad,[4] aimed at rendering complicated concepts more accessible through graphic visual forms.

---

[2]For example, the Water Footprint Network often reports its data, comparing the water footprints of North America, the world average and the African continent (Mekonnen and Hoekstra 2011).

[3]Recently in Italy, only the Turin Polytechnic ("Water in Food" Group) and the Milan Polytechnic have been active in research on virtual water and water footprinting, but two European projects have been set up which will keep the research going over the next years—the EURO-AGRIWAT project, coordinated by the University of Florence, and the ViWaN project (The global Virtual Water Network).

[4]See the info-graphics available on the Website www.angelamorelli.com/water/.

3 Why Another Label?

The Wall Street Journal article of 17 February 2009, "*Yet another footprint to worry about: water*?", referring to water footprinting, ironically contested the proliferation of the many footprints that an individual should worry about (environmental, carbon and water). Instead, today, corporate water responsibility is now at the forefront and is widely studied in the private sector, by academics and NGOs.

So, why another label?

Taking into consideration what was previously argued about the importance of assessing the environmental aspects linked to the production methods of goods we consume, this chapter presents a label prototype which aims to inform consumers about the water sustainability level of that product. Studies regarding water footprinting and virtual water have raised the awareness that not all uses of water are the same and that the use of the same quantities of water in different contexts can result in quite different consequences (Hoekstra and Mekonnen 2012; Rodriguez Cabellos 2011[5]; Ridoutt et al. 2009; Renault 2002). In order to provide a more significant assessment of the impact on water resources, Table 1 proposes a sustainability scale of the virtual water content of agricultural products. The approach to water sustainability proposed here does not claim to be exhaustive, neither is it the first (Vanham and Bidoglio 2013; Milà i Canals et al. 2008). It introduces a framework of water typologies, by means of which it is possible to assess the environmental impact of their use for productive purposes. A lower environmental impact results in cases where the water used originates from:

- farming using as much green water as possible (preferred to blue water which has a higher opportunity-cost);
- countries rich in water (especially green water as because alternative uses in other sectors—industry; domestic—will not be hindered from the use of green water resources in agriculture);
- renewable sources and as little as possible from over-exploited or non-renewable groundwater.

Other considerations could be added to those of a water nature—for example, the impact on the land where erosion is concerned or pesticide use, and the impact on the air quality due to CO_2 emissions. To the consideration of environmental impacts, we could also add the socio-economic aspects. For example, socio-economic sustainability will be higher where the virtual water produced originates from production that promotes local employment and development, does not impoverish the original environment—and this does not exclude the use of local water resources by the local inhabitants—and does not cause any economic and environmental harm to the populations involved. In other words, the water used for irrigation must not reduce the water levels which could be used for other purposes

[5]For further details on a qualitative view of water footprinting, see the concept of the "extended water footprint"(Rodríguez Cabellos 2011).

Table 1 Virtual water scale: typologies, origin and related impacts on agri-food products, with examples of water label prototypes

Impact on water resources	**Type of water mainly used**	**Renewable water body/ over-exploited groundwater or non-renewable groundwater**	**Origin**	**Examples**	**Label preview**
Level I low impact	Green water	n/a*	Region plentiful or short in water (unimportant)	Wine from Umbria and sorghum from Kenya: non-irrigated crops. Maximum water sustainability. These crops are only rainwater fed and do not draw on the reserves of the regions of origin. Despite the 2 cases having different characteristics of scarcity (plentiful for one, scarcity for the other), this does not affect the water sustainability of the cultivation.	 The label shows the lowest impact on water resources, or the highest level of sustainability.**
Level II	Blue water with green water presence	Renewable	Region plentiful in water	Wheat from Ohio: produced with green and blue water, where the blue water is fully renewable, in a region which does not suffer from water scarcity.	
Level III	Blue water with green water presence	Non-renewable or over-exploited groundwater	Region plentiful in water	Maize from the Po Basin: despite the zone not suffering from water scarcity, the maize is irrigated by over-exploited fossil groundwater. Both blue and green water, given the plentiful rainwater.	
Level IV	Blue water	Renewable	Water-short region	Bananas from the Jordan Valley: with a renewable source and mainly irrigation water, therefore blue, given the scarce rainfall.	
Level V	Blue water	Non-renewable or over-exploited groundwater	Water-short region	Tomatoes from the Maghreb: mainly irrigated with blue water given the scarce rainfall in the desert, with non-renewable groundwater, e.g. the Nubian Sandstone Aquifer, in a water scarcity zone.	 The label shows a foodstuff's major water impact.

[a]The "non-renewable" typology is not applied, by definition, to green water

[b]For more detail on the potential and sustainability of non-irrigated agriculture, see Chatterton and Chatterton (1996), Gilmont et al. (2012)

source edited by the authors

with a higher return (industrial use) or more social benefits in areas where there is a serious shortage (sanitary services, domestic use). In general, virtual water is to be preferred where agricultural water use does not result in denying water to populations living in conditions of extreme scarcity. If we then wished to add considerations concerning working conditions, it would be ideal if the virtual water came from production which did not exploit the foreign workforce at a lower cost than the local one, as this creates labour market crises and unemployment for the small-scale farmers. Moreover, it would also be better to prefer cases where producing goods for export does not hinder small scale farming or generate lower incomes for local populations.

4 Virtual Water Scale Typologies and a Label Prototype

Table 1 presents a scale of the different virtual water typologies that we can find in an agri-food product, so that we can use it as a basis to create a prototype for a water label. In Table 1, we can distinguish not only the different types of water that we can use for agricultural purposes—that is, green water and blue water, which can be further broken down into renewable and fossil water or over-exploited groundwater—but also the context in which the water withdrawal occurred. For reasons of simplification, the zones of water origin have been subdivided into regions rich or short in water, also including in the first-category areas characterized by economic resource scarcity (for more details on different types of water scarcity see "Not all drops of water are the same" by Antonelli and Greco in this volume). For a more detailed "virtual water

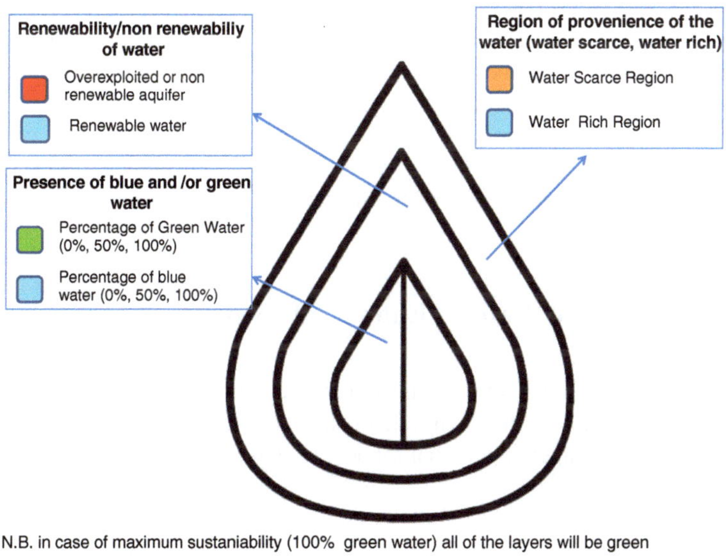

Fig. 1 A label prototype *source* the authors

scale", which also includes the climate zone of the product origin, the income level of the producer countries and the five types of different interaction between green and blue water, the reader can refer to Antonelli and Greco (2012).

On the basis of the different virtual water typologies in Table 1, we propose here a prototype for a water sustainability label (Fig. 1). This prototype is presented in the awareness that it is a proposal that can be improved upon and that, for a more complete and sophisticated design, it will be necessary to involve marketing experts, info-designers, producers providing information at source and all those players, both large and small, involved in the supply chain.

Grey water must also be considered of minor importance as in the productive processes for agricultural goods, it represents a minimal percentage (about 10 %) compared to those of blue and green water (Allan 2011), but this element could also be included in a later version of the label. Similarly to the discussion on grey water, the clear distinction between green and blue water (or the co-presence in equal parts) can be justified here by the need to simplify and communicate immediately what a label needs to get across. In fact, it is rare that the interactions between blue and green water are precisely 0, 50 or 100 %, and the intermediate solutions are those closer to reality. The label prototype proposed here attempts to quantify these interactions, distinguishing them into three possibilities (all green, all blue, or 50 and 50 %). The intermediate area, if red, indicates the use of non-renewable water or over-exploited groundwater; light blue indicates a source of blue renewable water (rivers, lakes, non-over-exploited or renewable groundwater). Another simplification on the label regards the distinction between the zones rich or short in water. As in the virtual water scale (Table 1), the criteria for physical and economic scarcity, defined in Antonelli and Greco's "Not all drops of water are the same", were encompassed in the more immediate category of "water poverty", found in the outer layer in orange. As far as the choice of colours is concerned, wishing to use warm colours to communicate a sense of danger or non-sustainability, and the blue and green colours for sustainability, the choices were the following: concerning only cases where there is a maximum water sustainability and an environmental impact of virtual water equal to 0 (that is, cases where non-irrigated farming is practised) to be able to communicate this solution as the best of all of them, we have used an exception to the established graphic norm, and we have coloured the whole drop green.

5 How to Become Aware "Eaters" of Water

People are different for what they believe, for their actions and for what they decide to choose. The voices of the environmentalists and the NGOs, according to Tony Allan,[6] provide an immense potential for actions that have not been exploited where

[6]Presentation on "Water and food security: which solutions?" during the World Water Week in Stockholm (2011).

governments and the market are concerned. The NGOs could promote and create common shared values on the issues of water through cooperating with and influencing the big international brands. They could even embark on campaigns involving all the food supply chain, promoting a sustainable production for all foodstuffs, if their potential were fully exploited. Knowing what impact the sale of a certain good containing virtual water has on local communities could be included in a new water ethic and become part of a new awareness of this common good, which Fantini defines as "the moral economy of water" (Fantini 2011). Therefore, it is quite a difficult job to make consumers of virtual water aware of something they do not see, do not know because they are not able to link their reality to that of the food origin. Being able to make the invisible visible and link the two realities—that of the consumer with that of the producer—in order to protect and safeguard a certain area and its inhabitants' rights, even if that area is far away and unknown, is what this study proposes to do, at least, as a first step in spreading information. The objective will be fully achieved when we stop speaking of water consumers and begin speaking of global *citizens* who protect a global resource, from a both local and global perspective. This would be a valid approach to initiate a discussion on awareness. Besides an approach of environmental education and work with the private sector, foodstuff labelling is definitely a path to embark on, orienting consumer choice to foster a sustainable consumption. Despite the numerous labels we are all familiar with and used to in making our choices in supermarkets, we claim it is still possible to attain a common standard for product labelling, coordinated more at an international level rather than at a local or national one.

Since the first publication of this volume in 2013, many other relevant facts happened. In 2014 the new ISO standard for the water footprint was released. The Water Footprint entered a new phase and new elaboration standards (Hoekstra 2013, WFN 2014); many other case studies and national reports have been published. In Italy, authors' interactions with academia, agri-business sector and farmers made it possible to explore additional uncovered problematics.

The new context-based water footprint (Hoekstra et al. 2012) method[7], which is a very context-specific analysis, might apparently seem in contrast with the idea of a standardized and universal water-label, but the two can be seen, in our view, rather as complementing elements. A water footprint analysis is at the same time feasible for measuring a company's performance in terms of water use, whereas a water label could be used as a synergetic tool, as a communication instrument to consumers and a reputational support for producers. In particular, regarding the prototype proposed in this chapter, the concept of "maximum sustainable water footprint level per river basin"[8] (Hoekstra et al. 2012) could be coupled with what is here called "non-renewable and overexploited aquifers". Extending this class to

[7]Context-based water footprint implements proper water footprint accounting in the context of a country or organization (Hoekstra 2013).

[8]In Hoekstra's global study on blue water scarcity per basin, it was assumed maximum sustainable blue water footprint levels at 20 per cent of natural runoff across river basins (Hoekstra et al. 2012).

"non renewable, overexploited ground and surface water" could potentially be explored. Another useful lesson from the Water Footprint literature is the one exploring the need for a reference for farmers regarding water using-processes (Hoekstra 2013). In the presented water-sustainability scale, a tomato will be always consuming irrigated (blue) water compared to a totally rain-fed durum grain of wheat. However, tomatoes are not usually considered rain-fed crops; therefore, not all crops can be assessed on the same scale. These considerations need to be addressed through further research aimed at defining a label-prototype. Unstructured interviews to farmers in Italy revealed concerns regarding the need to pay for obtaining a new label for their products (Greco 2014). This cost was considered as unbearable for the majority of them. Moreover, the water labelling of food products could potentially affect negatively the small farmers who generally promote zero-km production, green water use and non-intensive agriculture. Therefore, creating a labelling method which could end up being unfavourable to small producers or, worse, which could only favour big companies—more likely able to afford the assessment process—could be a fatal error. Another criticism emerged from exchanges with academia (Greco and Antonelli 2013). The promotion of green water at any cost could cause as well very negative consequences in terms of land pollution and soil erosion. This is true even when producing rain-fed crops or when using a very large amount of green water compared to blue water. This criticism is very well grounded and thus reinforces the idea that a multiple food-label is to be promoted. In this sense, it would be advisable to produce the most complete label type, containing the following sub-categories: water, air, land, human factors (labour conditions and impacts of the production site on local populations, encompassing conflicts and human rights infringements). Essentially, this book only covers one area of an ideal labelling effort. This proposal is also in line with what Hoekstra calls the "scenario n.2 of water labelling worlds: a world in which many products are not sustainable" and, however, "all products have an internationally standardized sustainability label that shows the overall performance of the product" (Hoekstra 2013:196).

Finally, in 2014 the International Organization for Standardization (ISO) developed the International Standard "ISO 14046, Environmental management—Water footprint—Principles, requirements and guidelines", which is intended to provide decision-makers in industry, government and non-governmental organizations with a means to estimate the potential impact of water use and pollution, based on life-cycle assessment (ISO 2014a). According to the ISO 14046 "a water footprint assessment alone is insufficient to be used to describe the overall potential environmental impacts of products, processes or organizations" (ISO 2014b). This claim is consequentially reinforcing the principle according to which quantitative assessment cannot prescind from being coupled with qualitative assessment, if the final aim is that of assessing products' environmental impact. Thus, also in light of the recent development of the ISO 14046, our proposal for a qualitative water labelling prototype can still be considered a valid starting point for further elaboration and inclusion as a complementary tool of analysis.

6 From Consumers to Active Citizens

The first step towards change is transforming ourselves from unaware consumers to active citizens, shrugging off what is a rather simplistic name confining us to being identified only by what we consume. Being informed about the origin of our food, choosing the short supply chain, or pasture-grazed meat or seasonal vegetables, re-socializing water and discovering local values and impacts would be the first aim of this great common effort. And not only, this effort would also be a first great step in fighting virtual water waste which, let us not forget, is true water in the respective countries of water source origin. Virtual water is invisible just as the mechanisms it works with are invisible. Bringing virtual water out into the open and making it visible to all is a way of raising common awareness to a first level, the essential step for a common future action against the waste of the most precious resource we have.

Bibliography

Allan, A.J. (2011). Water and food security: which solutions? In *Presentation in the occasion of the UN Water Seminar World Water Day 2012 coordinated by FAO for UN-water members*. Sept 24, Stockholm, World Water Week. https://www.yumpu.com/en/document/view/10688462/water-and-food-security-which-solutions-tony-allan-kings-college-/3.

Allan, J. A. (2011). *Virtual water: Tackling the threat to our planet's most precious resource*. London: IB Tauris.

Antonelli M., Greco F. (2012). Accounting for virtual water sustainability in food products: Presented at the "Food Security in Dry Lands" conference held in Doha (Qatar), Nov 14–15, http://slidesha.re/X8hsjU.

Boström, M., & Klintman, M. (2008). *Ecostandards, product labeling and green consumerism*. Basingstoke: Palgrave Macmillan.

Chapagain, A., & Orr, S. (2008). UK water footprint: The impact of the UK's food and fibre consumption on global water resources, WWF-UK, Godalming, 1, 31–33.

Chatterton L., & Chatterton B. (1996). *Sustainable dryland farming: Combining farmer innovation and medic pasture in a Mediterranean climate*. Cambridge University Press.

Crivellaro, M., Vecchiato, G., & Scalco, F. (2012). *Sostenibilità e rischio ecologico*. Guida all'integrazione degli strumenti di comunicazione ambientale: Libreria Universitaria, Padua.

European Commission (2011). *Innovation for a sustainable future—Roadmap for eco-innovation (Eco-AP)*, Commission communication to the European parliament, the council, European economic and social committee and committee for the regions, Brussels, 15.12.2011 COM (2011) 899 final.

Fantini, E. (2011). *Acqua privatizzata?*. Cittadella Editrice, Assisi: Economia politica e morale.

Gilmont M., & Antonelli M., Greco F. (2012), A development pathway to optimise sustainability of water investment and minimise social cost: *Conference paper presented in the "Food Security in Dry Lands" conference in Doha, November 14–15, 2012*.

Greco F. (2014). *Unstructured interviews to Italian small size and family farmers in the High Tiber Valley in Italy*. Unpublished.

Greco, F. & Antonelli, M. (2013). Accounting for virtual water in food products: a tool for business and consumers. In *The proceedings of seminar "What colour is your water? A critical review of blue, green and other 'waters'*. University of East Anglia- Institution of Civil Engineers (ICE) Westminster: London, SW1P 3AA. Retrieved February 22, 2013, from https://www.uea.ac.uk/documents/40159/2682184/antonelli-greco-virtual-water-in-food-products/cba30ef7-c970-454a-9088-bace1b7e3cc7.

Hoekstra, A. Y. (2013). *The water footprint of modern consumer society*. Routledge.

Hoekstra A. Y., Chapagain A. K., Aldaya M. M., & Mekonnen M. M. (2011). *The water footprint assessment manual: Setting the global standard*, Earthscan.

Hoekstra A. Y., & Mekonnen M. M. (2012). The water footprint of humanity. In: *Proceedings of the national academy of sciences*, Vol. 109, no. 9.

Hoekstra, A. Y., Mekonnen, M. M., Chapagain, A. K., Mathews, R. E. & Richter, B. D. (2012). Global monthly water scarcity: Blue water footprints versus blue water availability. *PLoS ONE, 7*(2), e32688.

ISO (2014a). *ISO 14046 Briefing note*. Retrieved Nov 13, 2014 from www.iso.org/iso/iso14046_briefing_note.pdf.

ISO (2014b). Retrieved March 8, 2015 from https://www.iso.org/obp/ui/#iso:std:iso:14046:ed-1:v1:en

Mekonnen M. M., & Hoekstra A. Y. (2011). National water footprint accounts: The green, blue and grey water footprint of production and consumption. *Value of water research report series no. 50*, UNESCO-IHE.

Milà i Canals, L. et al. (2008): *6th international conference on life cycle assessment in the agri-food sector: Proceedings. Towards a sustainable management of the food chain, Zurich, Switzerland. November 12–14.*

Renault, D. (2002). Value of virtual water in food: Principles and virtues: Paper presented in the workshop on virtual water trade, UNESCO-IHE. December 12–13, 2002.

Ridoutt, B. G., Juliano, P., Sanguansri, P., & Sellahewa, J. (2009). *6th international conference on life cycle assessment in the agri-food sector: Proceedings. Consumptive Water Use Associated with Food Waste: Case Study of Fresh Mango in Australia, Zurich, Switzerland, November12–14*, Nemecek T. & Gaillard G. (ed.). Agroscope Reckenholz-Tänikon Research Station ART, June 2009.

Rodríguez Cabellos J. A, (2011). *Usefulness of the water footprint for water management and planning in the Guadiana River Basin. 3. Seminar of the Botin Foundation Water Observatory, Madrid, February 11, 2011.*

RPA, Cranfield University. (2011). *Assessment of the efficiency of the water footprinting approach and of the agricultural products and foodstuff labelling and certification schemes*, Executive Summary ENV.D.4/SER/2010/0051r, European Commission Directorate-General Environment http://bit.ly/WLl6lh.

Teisl, M. F. (2007). *Labelling strategies in environmental policy*. Aldershot: Ashgate.

Wales, A., Gorman, M., & Hope, D. (2010). *Big business, big responsibilities: From villains to visionaries*. Basingstoke, Palgrave Macmillan: How companies are tackling the world's greatest challenges.

WFN (2014). *The water footprint assessment tool*. Water Footprint Network 2014. Retrieved March 8, 2014 from http://www.waterfootprint.org/?page=files/waterfootprintassessmenttool.

WWF. (2009), *Investigating shared risk in water: Corporate engagement with the publicpolicyprocess*, HSBC http://bit.ly/Vh02Vh.

Virtual Water, H2O and the De-socialisation of Water—A Brief Anthropological Journey

Mauro Van Aken

1 Virtual Water and Water as a *Commons*

The concept of virtual water has been crucial in reintroducing the elements of water use and "productivity" into the interpretation of the economic and political aspects of contemporary processes. Describing the quantity and quality of water used in the production of food, goods and services in relocated networks of commerce has led to revealing the economic aspects of water deliberately hidden in production systems for decades. Virtual water is an empirical indicator of water consumption, and through a volumetric analysis, it can make visible how and why water is consumed and where it goes in the goods which incorporate it—it is a measure of the volume of water *embedded* in the most common things.

This new "measure" is precisely what has allowed geopolitical and economic issues connected to water production and the increasing water scarcity in many areas of the world to spread well beyond scientific circles. Water footprint analyses, as well as the translation of normal everyday goods into their value in terms of virtual water, have helped a mobilisation around water and started a public debate on an issue that would otherwise have been relegated to a world of technical knowledge, bureaucracy, national and international policies that are often exoteric, or private, as they are reserved for an exclusive technical audience or reduced to a mere economic affair. This indicator has contributed to making water once again a public issue, bringing it back again into the arena of political debate regarding the neoliberal trends that increasingly see water as just an excellent form of investment. It is because the interests and profits in the water sector are huge—and since it tends to be a sector on its own, completely cut off from the rest of society—in the south and north of the world that there is now a widespread and global process where citizens (those living in marginal areas, in refugee camps, in slums or shantytowns

M. Van Aken (✉)
Cultural Anthropology, University of Milan-Bicocca, Milan, Italy
e-mail: mauro.vanaken@unimib.it

© Springer International Publishing Switzerland 2015

M. Antonelli and F. Greco (eds.), *The Water We Eat*,
Springer Water, DOI 10.1007/978-3-319-16393-2_9

that have sprung up in the great megalopolises of the south of the world, or the farmers who must irrigate their fields, for example) are turning into good "water customers", with all the resistance and contradictions that this entails.

Over the past decades, this progressive de-socialisation of environmental resources has resulted in making water a private issue, and it is only recently, through mobilisations and campaigns, that the public dimension has re-emerged (in Italy, with referendums) focusing on the links between virtual water and consumption models, lifestyles and "good practices", or criticism of increasingly water-consuming development models. At the same time, these events have tried to avoid relegating the debates on water scarcity and growing competition to a merely environmental cause, bringing them back into public economic and social life.

These are key aspects in the context of the progressive de-socialisation of water, which characterised water development models—but often "water consuming"—in the south of the world as much as in Italy. In short, virtual water has proved to be a good economic analysis tool, but also a medium for communicating and disseminating information, highlighting the otherwise hidden connections between hydrological cycle sustainability and the trends in agricultural product exportation, demonstrating the place of water in agricultural policies and its inherent political importance.

This has, however, also revealed the water interdependence of nations and the types of imports and exports of the water disguised in goods, unveiling an otherwise hidden aspect of those goods—an element all the more important in the context of reduced food security but also of the declining water autonomy in many situations in the south and, increasingly, in the north of the world.

The study of the relations between traditional farming and water in other cultures, where anthropology has played an important and historical role, has shown the importance today of understanding water—its use and its abuse—not only in the economic processes but also within their social and cultural dynamics. It is not a coincidence that water has become, and not only in Italy, the symbol of "common goods". In fact, water, in different ways, has a social and cultural life, which has mainly remained hidden. And cultures are strongly linked to water, as lifeblood, but also as social and cultural *medium* since antiquity.

The study of the relations between water and other cultures has highlighted the importance and relevance to understand the complexity of *commons* as communal management systems of natural resources: a complex whole not to be idealised, frozen in time (in the case of contexts in the south of the world) or in opposition to our management methods, but as historical institutions, cultural, productive and moral systems, technical and symbolic, and central to the use and, especially, the sharing of water, the most *relational* good. Cultures, also in Italy, have always named measured and defined water, or better "waters", with cultural models that are not at all in contrast with, but different to H2O. The mere measurement of water is not enough to understand what a *common* is, how it works and what its environmental, social or moral relevance is. And in many contexts of the south of the world, it is not enough to introduce new analysis tools, if the local knowledge and management systems regarding water are not also taken into account. Not only can measures of water be very different and involve many different cultural worlds, but also many aspects of water cannot be measured, as they are a part of the social and cultural dynamics and institutions which are non-measurable and, therefore, require other tools of understanding.

For example, in the Jordan Valley, the cradle of irrigation practices and strategies for coping with an arid environment, two local practices were central to managing water in the past by the nomadic pastoral desert dwellers (*Bedu*) or peasants (*Fellah*). These two practices were targeted at the very beginning of the British mandate period as obstacles to be overcome for the sake of modernisation. The first was the idea of tribal land, or *dirah*, and the second, the system of community rotation of agricultural land linked to available water sources, or *musha*[1]: practices that had traditionally established the types of ownership, management systems, roles of authority and the cooperative systems concerning water. The colonial water modernisation projects identified these two diversified and dynamic management systems as obstacles to be overcome and replaced them through new knowledge and centralised exogenous institutions.

Here, as in other colonial contexts, water measures were always a main area of confrontation with local cultures. In the Valley, the local populations not only used to think, but still persist in thinking today of water and its measuring in local distribution in relation to the ancient idea of *dor,* or irrigation turn in temporal frame. We could better translate this unit of measurement as a "social time"—a water turn is thought of as the water which flows in 4, 8 or 12 h, involving many variables depending on the season, the crops and the water availability. It is a social time because the person receiving the water turn is the male representative of an extended family or of a lineage who owns a piece of irrigated land. His measurement of water is recognised within a system of relations based on family solidarity and potential exchange. Moreover, the water time is social because it is flexible: if a farmer has more water than he actually needs, he can exchange it, divert it to his neighbour and vice versa; if his plants are suffering from water stress, he will be able to find a turn through the existing reciprocal relations.

The *dor* as a local unit of measurement is in fact a relational measurement, negotiable within the existing social and political system, while H2O, as cubic metres/pressure, is a rigid measurement, external to local relations, related to the water administration. The exogenous measurement, quantifiable and scientific, desocialises water from the local context by linking it to the needs, language and priorities of the administrative system. The local measurements of water, on which the ancient irrigation systems of these lands were based, allow for a "space of manouvre" in a context which has become increasingly more difficult due to an increase of water scarcity and competition. The *dor* is tied to local practices, an integral part of the social and political context, readily adaptable to the variability of territories, seasons and local needs. Therefore, speaking of *dor* recalls another

[1]These two historical experiences are not limited to the small area today defined as the nation of Jordan, but involve the economic and social history of many parts of the Middle Eastern region.

social world of kinship, exchange systems, patterns of reciprocity and the idea of *dirah* as tribal land. It also recalls an idea ecological constraint and an intimate knowledge of seasonal water variability.

The idea itself of water "scarcity" had never had a translation or relevance in these regions and was only adopted in the last decades as an exogenous idea of water modernisation. And this is not because of some obscure traditionalism in local knowledge. Water was not "scarce" because it was not thought of as "plentiful", but finite, and defined specifically and environmentally. In local terms and to orient the economic, agricultural and nomadic customs, water was thought of as "unpredictable" and "variable", two basic ideas in recognising the specific environmental conditions and the resources' limits. The social responses and the economic coping strategies had to be flexible in facing these dynamics, adjusting to the unpredictability and flows of water, deliberately choosing not to introduce any centralised or rigid systems for such a variable and fragile environment. The periods of *jafaf* (drought), a concept which was instead used, were recurring, for example, in the 1930s and 1940s, or at the end of the 1990s in the last century, and different traditional and economic coping strategies were consequently developed: increased mobility and transhumance practices (highly restricted today due to national borders), reduced pasturage activity, temporary employment in paid agricultural work, in sharecropping or in trading activities linked to nomadism, such as smuggling. All these strategies to reduce the economic and social risks were often limited with the introduction of state centralised systems. Scarcity, the main enemy to be fought today, has come about through a process of "construction of scarcity" in this valley. And a key factor in creating this scarcity over the past decades has been the increased cultivation for global winter markets of Mediterranean citrus groves, tropical banana plantations and tomato and cucumber crops in an arid and semi-arid environment.

These other *commons* are other stories of the interaction between societies and water, and more generally, the environment. It is not simply the fact of water being a common good (a term that today contains a number of meanings, uses and abuses and manipulations and unclear rhetoric), but the actual cultural aspects of water have defined different patterns of *commons*. This is because the social institutions involved in water distribution (always the most complex, being the key element concerning inclusion and exclusion, or equality and inequality), the roles of authority, the patterns of local water knowledge, obligations and prohibitions, social networks linked to farming water use, solidarity and cooperation networks, techniques, ideas of morality and deep-rooted symbolic meanings, all make up what we call *commons*, in order to grasp the "cultural diversity of water". Not water in itself, but the close relationship between society and the environment, between the communities and water, between culture and nature, and their co-production and interrelations have all created *commons*, often becoming the main target for water modernisation, with a specific intensity from the 1960s.

There have been many ways to relate to water, to work with water, to measure and define it, relations which have created complex social and political organisations. By providing some examples of this diversity, we can better understand the hidden meanings of our H2O, "our" translation of water which has led to a recent and contemporary management system, and which has defined an explicit way of maintaining a distance from water (just as from the wider environment), with contradictions in terms of water crises, development models and the emergent scarcity, also taking place in Italy.

The patterns of socialising water, of attribution of meaning, the social practices and techniques in relating to water have resulted in an incredible diversity and inventiveness among multiple cultures. Consequently, it would be more useful to begin speaking about "waters". This is not only because of the diversity and variety of water, which every culture, even more so in arid situations, has known and used, but more for the many different types of relations existing between societies and water in different contexts. By this, we mean the diversity of cultural models in using, thinking and creating social relations with and by means of water which, at the same time, make up different, but not necessarily conflicting, ideas of nature, of relations between culture and ecology, between water, equity and power. Therefore, it is important to measure virtual water in order to rethink what change involves, but, as well, it is necessary to reintroduce the understanding of water in culture and within the environment, to learn from other contexts, from other "water experts", and reflect on our ideas, conceptions and practices around water.

The local waters, in many parts of the world, have increasingly become a global business. The commodification of water, the rural development models based on intensive irrigation, the deterritorialisation of water and its radical disassociation from ecological, social and cultural systems are issues which are at the top of political agendas, but they are also the drivers for new types of political mobilisation and social conflict. Despite this global "visibility", water is often studied from a technical and economic perspective or from geopolitical viewpoint which, however, removes it from a sense of perspective, from a deep understanding and from its link with the local context and the cultural and social aspects which water has always played a part in. We now wish to focus on this aspect, and only a long journey can allow us to better understand the contradictions in and our conceptions of H2O.

1.1 H2O and the Abstractions of Water

Even though when we think of water, a tap immediately comes to mind—domestic and drinking water—it is irrigated agriculture, and even more so the water-consuming systems used in intensive and industrial agriculture, which is the main contributing factor to water stress in many situations in the south of the world, although this is also the case in the Po River Valley in Italy. Many cultures based on irrigation systems created very complex "social waterworlds", political organisations and cultural and agricultural systems that were, however, rendered invisible throughout the history of water development.[2] At the same time, there were very few public debates (also in Italy) and mobilisations in the south in relation to irrigation privatisation or to the power dynamics in irrigation practices.

Although traditionally delegated to the domain of technical knowledge and to quantitative perspectives in our culture, irrigation has a very complex "social life". Distributing water to fields has been an intensive activity involving notions of family and of work systems, cultural perceptions of the environment and patterns of authority, ideas about the ownership of water, and also social and technical practices in relation to irrigation. Water has a social life because it is a process where the environment interacts with social and cultural dynamics. Moreover, irrigation involves a complex application of knowledge regarding the environment, an important agricultural activity which is, however, facing a critical agrarian crisis and a growing emergency in water scarcity.

The invention of water as H2O, and the social world to which it is connected, can be traced back to colonial times as the control of this resource was a driving force in conquering vast areas of land and was instrumental in controlling populations in order to maintain the stability of colonial empires. Modernising water was inevitably a political and pedagogic project to modernise "others", even more important in agricultural contexts, and introducing modernisation was often, by means of water, a first step to censuring or ignoring the local knowledge systems and *commons*. Contemporary matters regarding scarcity, competition, local participation, opposition to technical changes and the political aspects regarding the "appropriation" of water in the south of the world are questions that actually originate from the colonial era. Therefore, the colonial experience was, to a large extent, a hydraulic experience. The water projects throughout the British Empire, even in their diversity, were, for example, defined as "civilising canals" (Gilmartin 1994), that is, hydraulic, territorial and irrigated works establishing colonial agriculture as the foundation for the stability of, firstly, the colonial empire and afterwards for that of the state. The importance of these water modernisation projects, along with the moral and political objectives, was based on the concept of the colonial encounter—to incorporate and domesticate new lands and new people with water modernisation as the foundation of the state and new central authority. In order to create agricultural (and irrigated) colonies, the "canals of modernisation", also based on adapting the pre-existing systems or inventing new ones, usually borrowed from local water knowledge and methods, when it served them, due to a need to control and to co-opt the local elite groups.

Irrigation, with its uses and abuses, is today at the centre of global agri-food production systems, of intensive agriculture and of crucial issues regarding food security, such as that of water autonomy. This is due to the fact that in contexts where there is an increasing loss of water autonomy and growing competition, irrigation uses a very high percentage of the water available and inevitably is seen

[2] See, for a broader anthropological view on the relationships between water and culture, Van Aken (2012).

as a strategic sector. Irrigation was at the centre of the agrarian revolutions, such as the Green Revolution of the 1970s, which invested in the "irrigationist" myths and the modernisation of the water "sector" at the expense, for example, of other farming and agricultural systems, such as the traditional rain-fed ones. Jordan is a showcase in this respect. Irrigation management in Jordan absorbs approximately 75 % of the annual available water, and given the forecasts for a lack of water autonomy by 2025, in recent years, water has passed from a sector of "rural development" to a one of "national security", opening up the way for emergency policies that often censor any possibility for public debate. The Middle East is today one of the areas of the world classified by international bodies as being subject to "water stress", and different countries are now looking at their limited water resources facing years of rigid and water-consuming development models, making irrigation the main issue regarding national sustainability. Today, at the heart of this "crisis in waterworlds", a fierce competition takes place among the different economic sectors, *in primis* between cities and irrigated rural areas, between national states regarding transnational waters, and particularly, among local communities.

We are now facing a crisis in "waterworlds" regarding the clear environmental and climatic contradictions of the development models, the crisis in the types of cooperation linked to water management and the spread of conflicts and "scarcity". And by "waterworlds", we mean the intense, multidimensional and cultural involvement, which all societies have enjoyed with this natural good. The water crisis has a long history. If at present time, water is one of the icons of globalisation process and of technical innovation already in the early colonial period irrigation models were already perceived as universal and "global". The crisis in the waterworlds is, therefore, also a crisis in the water-consuming development models, which have often been imposed on different local contexts. This crisis has come about due to a double abstraction: the exportation, by means of development policies, of models of water management that ignored, or specifically aimed at replacing, local management systems in other cultural and environmental contexts. These often transcended the historical multidimensionality and relationality of water. At the same time, a second abstraction cut across the local relations between culture and water—institutions, labour, technical and ritual systems, moral systems rooted in ancient stories linked to irrigation were disrupted or relegated to "traditions" to replace. In short, we exported an idea of water abstracted as H2O—from its social and cultural multidimensionality, for "virtual peoples"—abstracted from their cultural and environmental histories, from their names and their water techniques. A fundamental disconnection took place: the most relational resource in nature became sectoralised, consigned to a technical and economic sphere, removed from its social and cultural dimension as if it was a purely "technical" question unrelated from society. Consider the knowledge and culture of rivers, the local aqueducts constructed in the hill and mountain areas which make up the backbone of Italy, or the water and rural network that built the Po River basin, or the local water management systems of farmers in landslide risk areas. Thus, we have witnessed both the disappearance of different types of "water diversity" which were at the centre of many cultures, as much as the cultural and social patterns that are so important in the methods used by many communities to make their lands productive, to construct irrigation networks and social systems, to build landscapes, to "design" mountains' areas or face situations of drought, as well as the dangers and devastation that an over-abundance of water can cause.

1.2 The De-socialisation of Water/H2O as a Process of Modernisation

We take water for granted, in terms of H2O. We have taken for granted that it was infinite, based on the infinite potential of technology, underlying the "irrigationist" myths (Adams 1991) of industrial agriculture, nowadays in crisis because of these constraints and contradictions. It has been taken for granted that an irrigation system could be understood as detached from its social and cultural dynamics. We have taken it for granted that water has become a "stuff" (Illich 1985), an "object" to be exploited, while many other cultures have recognised its subjectivity and its active role throughout history. Moreover, it has been especially taken for granted that this idea of water, and its cultural relation developed in the last century in the "developed" world, was, even more than other models, a universal and transcultural given and, consequently, could be discretely exported to different corners of the world.

Having a tap in our home represents the main experience of how we relate to water, and also of what we consider a modern home and city to be. The main network, which provides life to our homes is a mostly unknown and hidden system —it is underground. We pretend that water as 'pure nature' enters our home, but at the same time, we hide the production and distribution processes of that water, from which we have been de-socialised, and, therefore, wish to remain unknown. We have moved water closer to the protected space of the modern home but we have distanced ourselves, in our direct involvement, from this good of which we ignore all—its origins, its journeys, its production and distribution systems, and its level of purity/impurity. This is all because we have delegated its management to an administrative apparatus and to a state service as if it was a discrete process.

To some extent, a commodification of water has already occurred here—its disconnection from other forms of direct involvement, its being conditioned by expert knowledge, by hidden productive systems and technological networks, which, however, must remain far from our sight and comprehension, often esoteric and distant, or merely "technical". We have transformed water from being contiguous to "hidden", invisible in its journeys and in its "production" (management systems and technical systems), a symbol of the other "nature", which must remain quite separate from the social life of the city. In fact, it is not so strange that the social changes that took place due to the water modernisation in the south of the world were called "tap revolutions": an urban model of water has been exported to, or imposed upon, rural areas, taking for "granted" the water contained in food production, which the concept of virtual water is, instead, trying to bring to light.

The translation of water into H2O has often resulted in cultural disorientation and loss of "sense of place". Water flows have been delocalised, local springs pumped to provide drinking water to distant areas, sacred places or political systems associated with water have been swept away, and people stripped of their skills and knowledge and reduced to the role of a "modern" irrigator, becoming an imitation of exogenous development models. In short, the local water, land and community networks have been disconnected in order to set up new larger networks and with new political players. This has not occurred on a *tabula rasa* of different contexts in the South, but on the contrary, it has occurred in areas where irrigation and resource management patterns had a long history that recognised its obvious limited and finite availability, and presented other cultural models, other labour systems and social networks built through water. H2O has been de-socialised, but it is also de-socialising. In the encounter with other traditional ways in relating to water, it has demanded the removal and censure of local patterns of sociality seen as useless or non-productive or, at the most, as an important part of local folklore or an "immaterial heritage", but not as an active element of contemporary agriculture.

2 Other Ideas and Water Networks: The Balinese "Subaks"

At this point, it would be useful to consider a case studied of a common in anthropology that was at the centre of the debate on development and of water experts in the 1980s in Bali. It is the case of the *subaks*, an historical *common* greatly misunderstood in Western perspectives, based on irrigation, but involving also political and cultural systems which, for centuries, had allowed for the intensive irrigation of terraced rice paddies carved into mountain sides (a landscape which has become a touristic resource for its scenery). This pattern of water management was autonomous from the state, decentralised with a public knowledge of the water system radically different from ours and from our H2O management—where for example, ritual and religious systems have been the key elements in technical and hydraulic organisation.

Geertz (1980), one of the fathers of contemporary anthropology, tackled the relational complexity between cultures and the environment, with a study on water in a comparison of "traditional" irrigation systems in a region of southeastern Bali and in a semi-arid context in the centre-east of Morocco. Geertz shows how the hydraulic organisation in both contexts involved an intricate social world, which shapes the environment and at the same time is influenced by it. Moreover, an irrigation system is presented as a social and cultural world, a *commons,* as we have defined here: the patterns of technical and social organisation are tied to different ideas of social relations, of political dynamics, as well as of agricultural economy.

Bali has an average rainfall of between 2000 and 3500 mm annually, with little difference over the years, the temperature remaining regular and constant. This stable and main water source has allowed for rice farming over the centuries, in a countryside that has been shaped, sculpted and terraced as a community project. Water in Bali is plentiful, but this has not deterred the development of a complex system, involving a controlled and regulated water distribution in order to avoid any inequality in distribution or any form of competition or the creation of water scarcity, something that often occurs in many contexts with apparently plentiful water resources (Waller 1994; Mehta 2001).

Given its economic importance, irrigation in Bali plays a "universal" social role, where an efficient, regulated and complex system exists—"an intricate hydraulic organisation" made up of "water neighbourhoods", which are not only technical subunit of irrigation but also social and political sub-units—called *subaks*. These make up the "community enterprises" closely tied to rituals and local religious authorities, social and very cooperative-oriented units recognised for their public but also extra-agricultural role. Water distribution is often the most socially complicated activity in irrigation practices because of its political and ecological aspects since it tackles the joint management of a common good. In Bali, this is coordinated by elaborate systems both technical and ritual, which is something we see as quite separate (technology as inevitably divorced from ritual). Knowledge of water here is also religious knowledge and the water "experts", those who have the knowledge of the distribution methods, are also identified in religious authorities.

The *subak* system has become a famous example of efficient hydraulic and agricultural management in relation to a social organisation that has been the focus of many studies (especially, Lansing 1991) since it went into deep crisis following the introduction of the Green Revolution beginning from the 1980s, a process which in many parts of the world undermined local water systems and *commons*, replacing them with new hybrid seed and irrigation techniques. The *subak* is an independent and autonomous system, and Geertz defines it as an "irrigation society"—a socially autonomous organisation solely dedicated to agricultural irrigation. A *subak* is made up of rice paddy terraces irrigated by a main canal bringing the water from a catchment weir higher up the mountain 5–10 km away and which is owned by the *subak*. The irrigated fields make up a communal land where each plot owner is automatically a member of the *subak*. It is a technical unit—its complexity often being the torment of many academics—but it is also a physical, social and religious unit. Each of these aspects is an integral part of the intensive rice production and the coordination of the collective work rhythms.

The main *subak* canal is divided by bamboo separators into secondary and tertiary canals with up to six/seven divisions reaching the cultivated terraces and forming an irrigation system down and along the mountains and valleys. Each of these forms a unit called a *tempek*, a quarter of a *subak*. The organisation of this preterraced canalisation is very complex, the subdividing of the water and the land is regulated, fixed and written down in the *subak* constitution on a piece of palm leaf, representing the whole system.

When the water reaches the terraces, it is subdivided into new sub-units, defining a "neighbourhood", so precise that Geertz also shows the equal and precise division of the water units in relation to the cultivated land units. This landscape is parcelled, measured and regulated reflecting an efficient and effective social organisation. Many farming activities are carried out individually but are rigorously collective where timing is concerned, while irrigation, instead, is strictly collective where the teams of workers are coordinated by the *subak* sub-unit officials.

A *subak* political chief coordinates the activities and is elected by the members and by a council. The latter, in turn, has the power to fine members for any violations of the common rules and elect the priest to preside over the rituals, which take place in the *subak* temples positioned on the mountain slopes at different altitudes, celebrations that signal the start of the agricultural and irrigation activities. This social system is a "bureaucracy in miniature", a highly organic social unit with a very strong moral feature. This "corporate service of a public utility" does not have to ensure a good water supply, given the very "wet" ecological conditions but, instead, focuses on the collective time and "rhythm" of water use which is highly important and rigorously controlled within the larger system, thanks to the technical experts—the *subak* chiefs—but even more the religious authorities who organise the farming rituals.

The crucial element, badly misunderstood or censured during the colonial period and then later by the national government and the World Bank, is that the rituals around the irrigation phases are also agricultural and irrigation methods. The rituals are a way of synchronising and coordinating the work with water in a communal landscape of rice paddies, conferring a "common rhythm to the farming work" or also, in some situations, establishing a large-scale coordination among the various *subaks* in different valleys (a network of decentralised systems).[3]

The big risk which this *common* seasonally faces is that the entire population of farmers on a mountain slop and valley use the water at the same time, causing an inevitable overuse of the water available which, particularly at the beginning of the season, is of strategic importance due to the high water demand in rice farming. The *subak* ritual system, therefore, "initiates" the work, firstly, higher up the mountain; then, progressively, the work moves down the mountain, in terms of time and altitude, towards the valley. This results in balancing the water demand depending on the varying needs of the rice paddies in the different seasons and on the farming operations, avoiding both seasonal overuse and underuse. An artificial environment that has, for centuries, permitted two/three annual harvests, guaranteeing an equal water distribution, always the most complicated factor in the relationship between water and society.

Lansing (1991) is an anthropologist who has continued to study these complex "artificial and engineered lands", which entered a critical period with the advent of

[3]The ritual tied to the *subak* is divided into nine main phases, the same in all the region: water opening, terrace opening, planting, purifying the water, feeding the gods with sacred water and other offerings, budding of the rice plants, the "yellowing" of the rice, harvest, placing the harvested rice in the granary.

the Green Revolution, following the new methods of water modernisation in the 1970s and with the radical change in water distribution and organisational methods. The author has queried why the functional role of the water temples had not become common knowledge for the development experts, from the World Bank to the national agronomist and water experts, and why had it remained misunderstood and censured to the extent that it has been replaced by new and much more simplified irrigation calendars linked to the introduction of hybrid rice seeds and new intensive farming models (Balinese agriculture had already been intensive with a double annual harvest). With the destructuring of this complex distribution system linked to the practices and ritual calendars of the water temples, the Green Revolution engendered a marked increase of pest infestations leading to a dramatic decline in rice productivity and the consequent increase in pesticide use, coupled with the introduction of the miraculous new hybrid seeds. Moreover, a second consequence was a "construction of scarcity", an abandoning of and competition in water distribution, terms that can appear to be paradoxical in a context characterised by Geertz as "wet" and highly plentiful in water.

In reality, the two consequences are interconnected. The introduction of the Green Revolution's intensive farming methods only partially absorbed the ancient ritual calendar of water distribution and which, let us remember, in an already intensively farmed system, allowed for an equal water distribution in many valleys connected to the same temporal system for an intensive, decentralised and networked use of water. This resulted in avoiding competition and scarcity in periods of peak water demand in the rice paddies. However, other rituals associated with the complex lunar calendars were, for example, abandoned and the synchronisation of water over an extensive water area left to disappear. Lansing[4] shows how the irrigation ritual system was not only a response to the water needs of the plants during the different seasons of growth, but it was, at the same time, the "creation" of an artificial ecosystem, an engineered landscape, spread over different valleys interconnected by water. This resulted in minimising the spread of pests, something that was, instead, reversed with the introduction of the Green Revolution. In brief, the system of water temples was and is, besides a water distribution system, also a complex anti-pest system as the regulation of the water in the fields at different altitudes removed pest-breeding habitats. However, the condition for this system of pest control was the synchronisation, defined by the rituals of the temple positioned hierarchically and at different altitudes, of the different farming operation cycles for the rice crops which required regulated quantities of water to actually minimise the risk of the pests spreading. By eliminating this synchronisation, the pests, instead, found a favourable breeding habitat in the wide terraces.

But how did this social and ecological system, as *commons* often are, become "invisible" in the eyes of the Indonesian engineers in the name of a new agricultural revolution? First and foremost, the social and ecological complexity of these

---

[4]See on Youtube the film taken from the video made by Lansing on this interdisciplinary and applied experience, "The Goddess and the Computer".

artificial systems was not only underestimated, but explicitly ignored. This was due to the perspective associated with water modernisation and its consequent changes —simplifying an ancient social and ecological system to a mere technical activity, where relations with the environment created over time were disrupted.

It is important to highlight how farming ritual and practice, temples and water management,were not only not in conflict with each other, but also were, on the contrary, overlapping: the "water temples define the institutional structure—the hierarchy of productive units—that manages the rice terraces as a productive system" (Lansing, 1987: 8).

Lansing proposes the notion of "ritual technology", where the irrigation and agricultural activity is not a series of technical procedures, but a "a meaningful series of interactions between social groups and the natural world. In this context, "the field rituals that accompany each stage of agricultural labor form a kind of commentary on the productive process. Moreover, the rituals of work in the fields may be "performative," in that they call forth particular social groups to engage in activities such as planting or harvesting. Agriculture, in short, is a social as well as a technical process, which is structured by the sequence of agricultural rites" (Lansing 1987: 10–11).

The water temples did not simply organise the water distribution, as Geertz had already noted: they used the water to create artificial ecosystems within a highly productive and common context as "moral communities", that is, interconnected by common moral and symbolic systems, involving prohibitions and obligations. It is clear how such a system could "work" without central state control, precisely because of the moral and political recognition of the *subak* as the local *common*. However, all of this was not translatable into forms of technical comprehension in which irrigation had been translated through development planning, where water was reduced to H2O; at the same time, the Green Revolution was destroying not a "balance" (*commons* are not balanced and unchangeable realities) but a complex and dynamic system of social and environmental relations. In order to translate to the experts of the World Bank or the Balinese Ministry for Agriculture this otherwise "invisible" reality, Lansing created, in collaboration with a computer ecology expert, a virtual simulation model for water management produced from a ritual and hierarchical scheme along the valleys. This allowed the development representatives to translate into "interpretable" data—since it is "measurable" and "quantifiable" just like our ideas and understanding of water—the relevance, complexity and "functionality" of these local systems. And this resulted in the state government investing once again in these social and ecological systems.[5]

Let us summarise here some of the consequences of this experience of the encounter (and ignorance) between different ideas of water and of "social life" which occurred in the rice paddies. First and foremost, Lansing shows how this system of temples was not historically tied to a centralised state entity but remained,

[5]See the video produced by this scientific, inter-disciplinary experience, but applied together with development questions http://www.youtube.com/watch?v=-cTiVpb5fdM

on the contrary, strongly autonomous despite the fragility of the Balinese state. Secondly, here, as in many other situations, we have a "ritualised technology": what had been made invisible—the role of the temples and the rituals—because it concerned the sacred and not the secular, became the domain of a new scientific authority, showing itself to be the central nerve of a social and technical system and of an agricultural system. Since the social life of irrigation was not linked to a single centralised situation, and the technical practices were interwoven with rituals, they have become elements that contradict the certainties of modern water knowledge, and as such, they risk becoming invisible, despite having shaped the Balinese landscape which is so frequently shown in tourist guidebook photos.

2.1 Water as a Cultural Medium

The Balinese case, as in other studies on the interaction between society, culture and water (Mosse 2003; Bernal 1997; Casciarri 2008; Chambers 1980), particularly in agricultural systems—and not its sectoralisation into an economic and technical resource—shows how water management is closely tied to social systems, to technical knowledge (also deeply social), a profound knowledge of a place and its environment, of political roles and networks. In short, within the flows of water runs a great deal of culture in complex and diversified patterns. And everything that cannot be translated into our idea of H2O becomes invisible if water management is only seen as a technical and economic event, simply based on a quantitative and volumetric analysis, which stands at the base of our water management methods. It is here that the communal, social and cultural system of water, which creates an historical and dynamic *common*, is preserved (or may fail). Therefore, water drives complex political and cultural relations, which are the lifeblood of many systems of water use. And first and foremost, these systems are "public" in the broadest sense. They are an active performance of the irrigating community (being also heterogenous and stratified), where thefts are visible, but also where the same community ties are created as much as the belonging to a common hydraulic network, which, at the same time, is a cultural and moral reality. And it is much easier to build large and new infrastructures and costly bureaucratic apparatus for water rather than to revitalise the communal management systems after they have been destroyed or made illegal, as has often occurred throughout the history of water modernisation.

An important element regarding the diversity of water compared to other environmental resources is clearly its very relational nature. In many cultural contexts, it appears to be impossible to understand water if it is not seen in relation to the land, the earth and the forests. In short, it is relational in exactly the same way we interpret ecosystems. However, central to this is its characteristic of being an important symbolic and social medium. In irrigated contexts, water interacts with the environment and culture, but it is also a medium between social systems, institutions and the solidarity patterns among social groups. In short, water mediates social and cultural relations and in irrigation systems, it has resulted in complex systems and social organisations. Water, in irrigation, as in our urban drinking water systems, mediates complex levels of social and political organisation, connecting the water infrastructures to the political apparatus, to the cultural and moral dimensions—from the type of ownership, the distribution rules or the value systems —in "social worlds" of water. These have been characterised in many different histories of irrigation by two important aspects: the flexibility and the negotiability of the appointed social systems. Instead, where the social negotiability of distribution networks is reduced, conflicts and competition often emerge, as now occurs in the new management systems, which have often strained the relational and negotiable character of water. And often, the loss of this negotiable character of many *commons* has disappeared because water has become invisible, hidden, isolated from society and removed from its social and environmental dimensions.

Water also has a relational character because it is a driver and a medium of cultural and political relations. In the very act of irrigating, religious, technical, ritual, political and environmental dynamics are often condensed into non-agricultural meanings going well beyond the irrigation aspect itself. Moreover, water is a central medium in the history of humankind and types of settlement; it is the basis for the first, even large-scale, social systems; it obliged populations, even quite different ones, to work and cooperate together. In short, it has made agricultural and political systems interdependent and interactive. Water is competition, but over all it underlies the diverse cooperative and also decentralised systems of different cultures throughout history.

Water was rarely thought of as merely something to be passively and silently managed, a "*stuff*" as Illich (1985) defines it, stripped of its material and symbolic multidimensionality, in order to make it "available" for human management. In other models of relationship between water and culture, water's "subjectivity", for its capacity to create and shape the environment and human history, was often recognised as fundamental, and not only in folkloric or sacral terms. In the irrigation practices and in the artificial systems that irrigation gave rise to in the different regional and farming contexts, the relationality of water has often been at the centre.

The monodimensionality to which water has been reduced in being measured in terms of H2O or in terms of particularly water-consuming development models is not something that helps us in any way to understand the complexity surrounding water which other cultures have dealt with. At the same time, it does not help (and even hides) in understanding our own production and social systems regarding water, the *hidden hydraulic aspects* of our societies. This intrinsically relational aspect between environments and cultures makes water one of the first social media. Water was valued and enhanced culturally as the mediator of many meanings, interconnecting social groups, regions and economic systems.

3 Water as *Commons*, Today

The issue of natural resource management has been influenced by the criticisms of Harding (1968), in his essay entitled "*The tragedy of commons*". In this interpretation, the author began with the premise that local systems were tragically unable to manage resources economically, effectively and sustainably, resulting in social and economic collapse. This view easily explained and justified the old evolutionary and moral prejudices regarding the technological backwardness of local knowledge in other cultures, tied to a presumed need to introduce private ownership and a centralised state management of resources so they could be managed rationally and effectively. This interpretation often legitimised the grabbing of resources in southern regions of the world and the censuring of local knowledge, seen as something backward or to be replaced.

The 2009 Nobel Prize Winner for Economy, Elinor Ostrom, became famous, first and foremost, for her work on the management of *commons*[6] regarding water, and her studies were developed specifically looking at the centrality and complexity of irrigation systems in the south of the world. According to Ostrom (1993), it is not water in itself which is the common good, but the political and social organisations involved with water, which are a challenge and an important potential in tackling the environmental issues and water scarcity of our contemporary world. Even if the patterns of common resource management seem to be dichotomously sandwiched between national and local management, and between private and "communal" ownership, in reality, these historical situations, far from being unchangeable and social fossils, have come once again to the attention of the development world for their capacity to adapt. They are also today a social force where climate change is concerned due to the flexibility and sustainability of decentralised management.

The Italian translation of commons as "*beni comuni*" (common goods) has often become today a *passepartout*, giving rise to confusion where, on the one hand, it has resulted in different types of social mobilisation (as in the Italian case), and on the other hand, it has lost its capability for social criticism and to understand the relational systems between water and society.[7] It is not for water *per sé* to be a *common*,[8] but rather the sociocultural interface of water, the reciprocity between nature and culture—in short, the management systems with all the rules, ideas of community and environment, roles and prohibitions.

In the frame of the different legal and economic definitions used in the Italian debate around water as common, let us recall the broadest meaning that anthropologists developed to understand local systems and knowledge, which focused

---

[6]We will keep here the English term *commons* as the translation of "common good" leads to too many different meanings in Italy.

[7]Only in the Italian scenario, different political actors and parties have adopted rhetorically this term for the more varied "common goods" with a loss of meaning.

[8]If not in the legal and economic definitions of common goods, but here we propose instead *commons* as a tool for understanding.

attention on the unavoidable sharing of water in a common network, on the relevance and dynamics of local management systems, on the "public" role in the wider meaning of *commons* because it actually reflects a moral and value system, but also a common "*savoir faire*" as it occurs in irrigation, so vital to farming systems. And therefore, water as *commons* does not coincide with the absence of hierarchies, with a nativist idea of communitarianism or with an homogenous community, of free and open access or lack of contradictions: commons can also "fail" due to the emergence of divergent interests, the arrival of new exogenous actors (as in the two cases studied), and the loss of sharing a common and moral water system.

At the same time, water as *common* highlights, as the diversity of cultural models has taught us, the importance and effectiveness of a communal management starting from the institutions, knowledge, methods, status and rules which have resided in and created an environment based on water and its network. Water has become a social institution whose performance, adjustment and dynamics are a part of the meaning and sense of a place and the meaning of "us", a public dimension recognising our active involvement with water and the environment. Instead, the definition of "communal", of what "community" means, becomes more difficult as the many asymmetrical interests at play have fragmented the meaning of local community in many contexts where water is concerned.

4 Conclusions: What are We Hiding in Water?

As we have seen, water involves many aspects in different cultures—its close relationship between the environment and culture, its multidimensionality, its strong imagery and symbolic elements together with its fundamental role in agricultural systems, its dynamics in capitalist incorporation and in other economic patterns. As it has occurred in our urban networks, water has become increasingly hidden, inaccessible and forgotten in underground systems as if it is actually no longer visible, as well as in rural systems transformed by agribusiness with underground pressurised water systems. Together with the loss of our closeness to water, the decisional levers have been hidden as much as the political aspect of these production systems exemplified by modern irrigation networks. The management of H2O has resulted in naturalising the political and cultural dynamics of change: if water is a technical issue, it is probably a subject of interest and work for technicians, consequently resulting in a denial of the political sphere, while the local population involved will continue to perceive it as something pertaining to social and power relations.

Furthermore, by removing water from its political sphere and delegating it to a technical world, our relations and the relations of others with water has been "naturalised": just as for tap water, or for the water used in the food we eat, we take naturalness for granted, forgetting, instead, the key social, cultural and political aspects. We have, therefore, abstracted the many values in the meanings which flow in water and which cannot be measured from a reductionist and managerial point of view. Thus, we have hidden to ourselves the radical changes that have occurred in our society and the contradictions between our society and the environment, based on censuring other cultural "waterworlds", the key element in water diversity.

Illich used in his important essay on water the term "waters of forgetfulness", a concealment and removal together increasingly relevant today as key aspects in understanding the relationship between humans and water: "Water throughout history has been perceived as the stuff which radiates purity: H2O is the new stuff, on whose purification human survival now depends. H2O and water have become opposite: H2O is a social creation of modern times, a resource that is scarce and that calls for a technical management. It is an observed fluid that has lost the ability to mirror the water of dreams" (Illich 1985, p. 76).

The transformation of water into "stuff", as Illich shows, or the many different meanings and social uses of water reduced to H2O, represented one of the most radical, and also global, changes in transforming the relationship between human beings and their environment. Reintroducing the flows of water into culture, within a local "social world", into the environment and the history of the different types of relations interrelated by water, such as *commons*, is an important factor in order to not "pretend" that water is infinite and distant from us, but, instead, we depart once again from our active involvement with water and with the environment.

Bibliography

Adams, W. M. (1991). Large scale irrigation in northern Nigeria: performance and ideology. *Transaction of the Institute of British Geographers, 16*(3), 287–300.

Bernal, V. (1997). Colonial moral economy and the discipline of development: The Gezira scheme and 'Modern' Sudan. *Current Anthropology, 12*(4), 447–479.

Casciarri, B. (2008). Du partage au clivage: marchandisation de l'eau et des rapports sociaux dans un village du Maroc présaharien (Tiraf, vallée du Dra), *Anthropologues et économistes face à la globalisation,* E. Baumann, L. Bazin, P. Ould-Ahemd, P. Phelinas, M. Selim, R. Sobel (a cura di), Parigi, L'Harmattan, pp. 87–127.

Chambers, R. (1980). Basic concept in the organization of irrigation. In E. W. Coward (a cura di) (Ed.), *Irrigation and agricultural development in Asia. Perspectives from the social sciences* (pp. 28–50). Ithaca & London: Cornell University Press.

Chapagain, A. K., & Tickner, D. (2012). Water footprint: Help or hindrance? *Water Alternatives, 5* (3), 563–581.

Geertz, C. (1980). *Organization of the Balinese Subak, Irrigation and agricultural development in Asia: Perspectives from the social sciences*. Ithaca & London: Cornell University Press.

Gilmartin, D. (1994). Scientific empire and imperial science: Colonialism and irrigation technology in the Indus basin. *The Journal of Asian Studies, 53*(4), 1127–1149.

Harding, G. (1968). The tragedy of commons. *Science, 162*, 1243–1248.

Id, (1991). *Priest and programmers technologies of power in the engineers landscape of Bali*. Princeton: Princeton University Press.

Illich, I. (1985). *H2O and the waters of forgetfullness*. The Dallas Institute of Humanities and Culture, Dallas.

Kaika, M., & Swyngendouw, E. (2000). Fetishizing the modern city: The phantasmagoria or urban technological networks. *International Journal of Urban and Regional Research, 24*(1), 120–138.

Lancaster, W., & Lancaster, F. (1999). *People, land and water in the Arab Middle East.* Amsterdam: Harvood Academic Publishers.

Lansing, J. S. (1987). Balinese 'water temples' and the management of irrigation. *American Anthropologist, 89*(2), 326–341.

Lansing, J. S. (1991). *Priest and programmers. Technologies of power in the engineers landscape of Bali.* Princeton: Princeton University Press.

Mehta, D. (2001). The manufacture of popular perception of scarcity: Dams and water-related narratives in Gujarat, India. *World Development, 29*(12), 2025–2041.

Mosse, D. (2003). *The rule of water. Statecraft, ecology and collective action in South India.* New Delhi: Oxford University Press.

Ostrom, E., & Gardner, R. (1993). Coping with asymmetries in the commons: Self-governing irrigation can work. *Journal of economic perspectives, 7*(4), 93–112.

Van Aken, M. (2012). *La diversità delle acque: Antropologia di un bene molto comune* Altravista, Lungavilla.

Ward, C. (2003). Acqua e comunità. Crisi idrica e responsabilità sociale, Milan, Elèuthera.

Waller, T. (1994). Expertise, elites and resource management reform: Resisting agricultural water conservation in California's Imperial valley. *Journal of Political Ecology, 1*, 13–42.

The Italian Mobilisation for Water as a Commons: Moral Economy and Virtual Water

Emanuele Fantini

In modern and industrialised societies, water—"conquered", purified, channelled and distributed by means of technology and increasingly complex systems—seems to have lost its central role, not to mention the sacred role it historically held in shaping cultures and civilisation. Yet, over the last years, in different cases, water has returned once again to the attention of the public and become the focus of debate, particularly concerning the controversial question of the privatisation of water services. In Italy, the public mobilisation on this issue reached its height during the referendum held on June 2011. Two of the four questions to be voted on had, in fact, the specific objective to block the privatisation of water services. Specifically, the first question called for the repeal of the rule for local authorities to select the water service manager by the end of 2011 through a call for tenders open to public, private and mixed companies. Instead, the second question proposed to eliminate the guaranteed return on invested capital percentage from the water service tariff based on a 7 % fixed official rate.

The result of the referendum for many was a surprise. From 1997, in the 24 referendums held, a quorum of 50 % plus one of all those with the right to vote, and necessary to be valid as a result, had never been reached. Instead, in June 2011, there was a turnout of 54.8 % of voters who voted "Yes" to all the questions with percentages of around 95 %. This success is even more significant if it is considered that in the months previous to the referendum, the political establishment and the media had boycotted and obscured the referendum issues (Carrozza 2012). Moreover, the two questions on water were the first in the history of the Italian Republic to be promoted not by parties, but by a wide social coalition which collected a record number of 1.4 million signatures on its petition.

These figures would appear to confirm the remarkable potential of water as an issue capable of arousing public interest, activating channels of political participation and encouraging citizen action. In this context, where, on a level of public

E. Fantini (✉)
Department of Cultures, Politics and Society, University of Turin, Turin, Italy
e-mail: emanuele.fantini@gmail.com

© Springer International Publishing Switzerland 2015

M. Antonelli and F. Greco (eds.), *The Water We Eat*,
Springer Water, DOI 10.1007/978-3-319-16393-2_10

opinion, can the concept of "virtual water" be introduced and what incentives can this idea offer in terms of redefining the content and subjects of the Italian mobilisation for water as a commons?

To answer this question, this chapter will firstly analyse the referendum's success in the light of the experience and history of the Italian water movement.[1] Secondly, the cultural, ethical and symbolic aspects of the Italian mobilisation for water as commons and its political implications, beginning from the idea of a "moral economy", will be looked at more closely. Finally, the motivations, the prospects and the challenges which the concept of "virtual water" potentially introduces regarding the Italian mobilisation for public water will be discussed, particularly concerning the redefinition of the epistemological approach to the question, of the issues at the centre of the debate, as well as the subjects involved.

1 The Italian Movement for Water from Its Origins to Post-Referendum

The success of the June 2011 referendum has been analysed with alternating different explanations. Those who have focused on the political aspect underlined the willingness of opposition party supporters and managers to deal the Berlusconi government a second blow, after the success of the centre-left in the May local elections, especially with the victory of Giuliano Pisapia in the Milan. Instead, others saw in the referendum result an almost anthropological mutation in the majority of the Italian population, enshrined in the rejection of the Berlusconian political and cultural model and the economic dictates of neoliberalism. Finally, some studies proposed a parallel with the protest of the Spanish *indignados* or the Arab Spring, celebrating the victory of the "people of the internet" over the political and media establishment and stressing the role played by the social networks in organising and motivating citizen participation.

Each of these studies offers interesting points in understanding the referendum result. However, when it comes to the two questions on water management, the size and success of the participation can be fully grasped only when situated in the historical trajectory—longer than 10 years—of the Italian water mobilisation, formally organised in the Italian Water Movements Forum. The Forum involves a wide coalition made up of the heirs of the so-called anti-globalisation movement, civic committees, environmentalists, international development NGOs, consumers associations, a few trade unions (particularly the CGIL public administration section), Catholic-based associations (ACLI, AGESCI, Pax Christi), dioceses, parishes and missionaries.

[1]The thinking on the mobilisation for public water in this contribution based on the results of a wider study on the Italian movement for water, presented in C. Carrozza's and E. Fantini's (edited), *Si scrive acqua. Attori, saperi e pratiche nel movimento italiano per l'acqua pubblica*, 2013

2 The Promotion of a New Water Culture

Two main inputs lie at the origins of the Italian water movement: on the one hand, the circulation of and later the official publication of the *Manifesto dell'acqua* of the economist Riccardo Petrella (University of Lovanio, Belgium; Petrella 2001) and the consequent setting up, in 2000, of the Italian Committee for the World Water Contract (CICMA), thanks to the work of "some international cooperation NGOs, some historical exponents of environmentalism and a minority element from the Communist Refoundation Party" (Molinari and Jampaglia 2010); and on the other hand, the allegations and appeals originating from the World Social Forum of Porto Alegre, inspiring groups such as the international network ATTAC, which participates in the anti-globalisation movement. In both cases, these inputs originate from outside of Italy, reporting on the negative repercussions of water service privatisation, specifically referring to cases in different cities in Latin America, Asia and Africa. The specific interests and political priorities of those who were more sensible to these messages contributed to shape the Italian water movement with an international and "Third Worldly" imprint which still remains today. The echo of such imprint was heard even during an exquisitely national event such as the referendum campaign. In this initial phase, the movement mainly focused on spreading information and awareness raising to promote a new water culture, inspired by the principles of water being a "fundamental human right" and "common good of humanity" and the need to address what is at stake globally concerning the issues of water resource access and management.

3 Support for Local Battles Against Privatisation

Since 2003, the awareness grew within the movement about the impossibility to restrict itself only to the information and cultural dimensions. In fact, in those years, some of the effects resulting from the introduction of the Italian national reform on water service management outlined in the Galli law [2] become much more evident. The aim of the reform was to overcome the fragmentation existing in the system, divided into more than 10,000 local operators, and to promote an industrial reorganisation in the sector, through introducing market logic and tools: the call for tenders as a tool to select the actor in charge of local water services management; contracts regulating relations among the key players; and the tariff as the main tool to finance the sector in name of the principle of the "full-cost recovery" (Carrozza 2011). The movement's action thus became linked to local disputes involving opposition to the delegation of the management of local water services to private or public–private companies, as well as to other local struggles regarding the safeguarding of water sources and groundwater against pollution or the bottled water

[2]Law 5 January 1994, n. 36 "Disposizioni in materia di risorse idriche".

business. In this phase, a second particularity of the Italian water movement emerged, namely the alliance with representatives of local entities (mayors, councillors, municipal and provincial advisors) and, especially, with those subjects (small and medium municipalities, town and provincial councils) excluded from the decision-making concerning the reorganisation of the water sector in Italy and from the alliances and merges among the former municipalised agencies, which now became multi-utility companies, controlled by the authorities of the main Italian cities. The alliance between the movement and the local authorities was recognised in 2008 with the setting up of the National Coordination of Local Authorities for Public Water.

4 The Struggle at National Level

The need to coordinate these local struggles and the willingness to highlight their broader relevance at the national level led to the creation in 2006 of the Italian Water Movements Forum (hereinafter the Forum). The Forum's first actions were the drawing up of a popular initiative bill to make water once again public, supported by a petition signed by a record of 406,000 citizens.[3] The bill was presented in parliament in 2007, but it immediately ran aground in the Chamber of Deputies' Environment Commission. Then, in the following years, the centre-right government led by Silvo Berlusconi adopted a series of measures, specifically with the so-called Ronchi decree [4] aimed at speeding up the involvement of private subjects in water service management, institutionalising the bid as main tool to select the management (public, private or mixed companies) and forcing local entities to hand over their participation quota to these companies (Massarutto 2010). In order to contest this move, the Forum chose to embark on the referendum road—initially, not without some perplexity, even within its own ranks. The preparation for the referendum and the following campaign provided the movement with the opportunity to accomplish another three steps. Firstly, it facilitated the involvement of some technical experts within the movement, especially lawyers, essential for tackling the maze of the legal system concerning water utility management, and specifically, to draw up the referendum questions. The involvement of these activist experts came about thanks to the meeting of the movement with a group of lawyers (Stefano Rodotà, Ugo Mattei, Alberto Lucarelli) who, in the previous years, had

[3]"*Principi per la tutela, il governo e la gestione pubblica delle acque e disposizioni per la ripubblicizzazione del servizio idrico*". The text for the bill is available in the appendix M. Bersani, *Come abbiamo vinto il referendum. Dalla battaglia per l'acqua pubblica alla democrazia dei beni comuni*, Edizioni Alegre, Rome, 2011.

[4]Law n. 166 of 20 November 2009, conversion with amendments of Decree-legislation n. 135 of 25 September 2009, resulting from urgent provisions to put into effect the EU obligations and the ruling of the European Court of Justice.

participated in a Ministry of Justice ad hoc Commission (Mattei et al. 2007, 2010), to draft a bill for the reform of the civil code regulations regarding public assets.[5]

Secondly, the referendum campaign resulted in widening the mobilisation to subjects and organisations which, up to that point, had not officially participated in the Forum. For example, the ACLI or consumers movement joined the referendum committee "Due SI per L'Acqua Bene Comune" (Two Yeses for Water as a Commons). Moreover, the campaign attracted and benefited from the spontaneous and creative participation of sport associations, scout groups, parishes, student organisations, artists and individual citizens who found in the fight for public water an opportunity for a civic commitment outside the traditional channels of political participation. The data from a survey conducted by the Cattaneo Institute showed how 16 % of voters in the referendum had also actively participated in the electoral campaign, and for 60 % of these people, it was their first experience in political activism (Fondazione Istituto Cattaneo 2011; Diamanti 2011).

5 The Post-Referendum: From the Battle for Water to Safeguarding the Commons

Finally, in the post-referendum period if, on the one hand, there occurred an inevitable decline in spontaneous and popular participation, on the other, the Forum's action was strengthened and legitimised as a national political subject, being the "custodian of the referendum result". The partial implementation of the referendum result and the increase in the involvement of private subjects and capital in the local public services pushed the Forum to come up with new initiatives in order to fully enforce the referendum results. Nationally, this involved a "civil disobedience" campaign which invited people to the self-reduction of the water bills, eliminating the percentage of the return on invested capital, as established in the second referendum question but still not implemented by the water utility policy-makers and managers. Locally, the Forum promoted the collection of signatures in support to resolutions calling for the municipal and provincial councils to make a commitment to transforming the S.P.A.s (share-holding companies) which managed the water services into public entities, based on the decision implemented in Naples by the Lawyer Alberto Lucarelli, appointed in the aftermath of the referendum as Councillor for the Commons by the mayor Luigi de Magistris.

The reference to the commons brings to mind another point regarding the post-referendum path of the Italian water movement. In the wake of the referendum's success, the notion of the commons has been increasingly associated with

[5]The Commission's work did not have any institutional follow-up. The acts from seminar of the presentation of the Commission's work are collected in the volume of U. Mattei, E. Reviglio, S. Rodotà (edited by), *I beni pubblici. Dal governo democratico dell'economia alla riforma del codice civile*, Accademia Nazionale dei Lincei, Rome 2010. Also see Id., *Invertire la rotta: idee per una riforma della proprietà pubblica*, il Mulino, Bologna 2007.

very different spheres (internet, culture, libraries, education, work) and adopted by political parities, research centres and other social movements, explicitly acknowledging of having been inspired the water mobilisation, such as the occupation of Teatro Valle in Rome, begun the day after the June 2011 referendum, or by the Italian Forum of the Movements for the Land and Landscapes.[6] The phenomenon is symptomatic of the aspiration by many to renew policy language and practices. However, the risk is to inflate the idea of the commons and consequently neutralise its potential for social criticism and political transformation (Rodotà 2012).

6 The Moral Economy of Water as a Commons in Italy

The Italian water mobilisation, by virtue of its resilience in space and time and its capacity to act at different scales, appears to be quite surprising in a period generally described in terms of a disaffection with politics and a dramatic decline in political participation and interest. Moreover, paradoxically, a survey by the Piepoli Institute carried out in 2012 revealed that 23 % of Italians were satisfied with the quality of water services and, indeed, 62 % were very satisfied, with an average score for the general service quality of 7.5 out of 10.[7] Similar figures were recorded both for the technical services (interruptions in water supply, water pressure) and for billing. Apart from some sensational cases, such as those in Aprilia and Latina, where the delegation of the water service management to a private–public company coincided with a sharp increase in water bills, or pathological cases, such as the inefficiencies in Sicily linked to Mafia infiltration in water management, the quality of the water services appeared to be satisfactory for most of the Italian population.

Therefore, the Italian water mobilisation cannot be reduced to a mere material claim, linked, for example, to the cost of the bill, which remains one of the lowest in Europe. In part, the mobilisation benefited from fears that the *status quo* could have been changed, with the risk of a worsening in quality and an increase in service costs popularly associated with privatisation processes. More generally, the Italian water mobilisation has ended up over the years taking on a political and cultural significance which goes beyond the mere question of local water service management. To understand this significance and the reasons behind the effectiveness

[6]Due to the movement for water concerning the experience of Teatro Valle Occupato cfr. Faris (2012), "Va in scena l'autogestione", *Internazionale*, 15 June.

The Italian Forum of movements for the land and landscapes officially presents itself as "a grouping of associations and citizens from all of Italy (on the Forum for Public Water model) which, maintaining the peculiarities of each subject, intends to pursue a single objective—to save the Italian landscape and regions from deregulation and illegal constructions" (quoted from the Forum website, http://www.salviamoilpaesaggio.it, latest retrieved on March, 4th, 2015).

[7]Istituto Piepoli, *Indagine nazionale sulla percezione del servizio idrico da parte dei consumatori*, June 2012, available at www.agcom.it.

and resilience of the mobilisation for public water, it could be useful to understand it in terms of "moral economy". The notion was elaborated by the historian Edward Thompson to analyse the riots during grain shortages and against the grain market in England in the 1800s. The objective was to avoid utilitarian explanations for collective action and to highlight what it expresses in terms of popular concepts of legitimacy and justice concerning community well-being, economic transactions and the role of the state in governing these (Thompson 2009).

7 The Symbolic and Moral Features of the Struggle

First of all, the notion of moral economy allows to highlight the symbolic and moral dimensions of the protest, in reacting to political choices perceived as an attack and violation of a pre-existing and superior values system. Specifically, national and local policies aimed at favouring the entry of private companies into water utility management have been seen as attempts to alienate the control of an essential resource, both for the symbolic construction of the local territory and for its material management. Moreover, these feelings have been strengthened by the climate of distrust in the idea of privatisation (Birdsall and Nellis 2002), exacerbated by the present global economic and financial crisis. The data from the Piepoli Institute confirmed these ideas—according to 91 % of the sample interviewed, "water is a universal resource which cannot be managed in a logic of profit", while 64 % believed that the management of water utilities must be entrusted to public bodies. The Italian water movement, similar to other international experiences, summed up these feelings in the effective slogan "water—a human right and commons". Furthermore, the fight against the privatisation of water services has ended up over the years taking on the aspect of a paradigmatic fight against the commercialisation of life in general and for defending democracy against market and financial interference. It is not random that the Forum's motto has become "You write it water, you read it democracy".

8 The Community Aspect of the Mobilisation

Secondly, the notion of moral economy invites one to look more carefully at the community aspect of the mobilisation. Because of its symbolic dimension, the issue of water management has become an opportunity and a tool to reimagine and reinvent political identities and belonging. In particular, the idea of water as a commons has contributed to supporting different perceptions regarding the nature and borders of the community where specific rights and duties are owed concerning the control and management of the resource. On the one hand, the reference to water as a "common good of humanity" (Petrella 2001), predominant in the early years of the movement, emphasises the international dimension of the protest, its cosmopolitan character and the associated practices of international cooperation. On the other hand, defending the "mayor's water", safeguarding the springs and public fountains, the ritual of local consortiums and independent aqueducts, the claims for a "either public or private, but local" management by subjects "with strong local roots" (Petrini 2008, 2010)—strengthened during the referendum campaign—encourage the stories and practices in rediscovering the local region. These are embedded in the tradition of localism, which is one of the classic cultural repertoires of Italian politics.

The reference to the commons also inspires the self-representation by the Italian water movement activists: it contributed to strengthening the sense of belonging to the movement and to shape its most peculiar practices of political participation—the refusal of a charismatic leadership, horizontal organisation and decisions taken based on consensus instead of a majority vote.

9 The Relation Between the Movement and the Institutions

Finally, the experience of the Italian water movement suggests some considerations on the relations between social movements and institutional political authorities. The active involvement within the movement of local authorities, administrators and representatives, the choice to adopt political actions based on constitutional and local devices (popular initiative laws, referendums, proposals to be voted on in municipal, provincial and regional councils) and the same proposal to transform the water service management from S.p.A. (private limited companies) into public agencies suggest the need to soften the traditional interpretation of mobilisation in terms of an expression of mistrust and refusal of the traditional institutions and procedures of representative democracy.

On the contrary, the above-mentioned practices seem to point to the willingness of the Italian water movement to play a substitute role regarding traditional channels of political participation, parties *in primis*. Moreover, in opposing "the privatisation of politics" through the delegation of traditionally public prerogatives to the market and private actors, the Italian water movement asks for a (re)assumption of responsibility on the part of the public authorities and for the introduction of spaces and devices for effective citizens participation.

All these elements together—the moral and symbolic aspects of the protest, the community dimension of the mobilisation and the request for public authorities to become once again more involved in the management of public assets—contribute to define the content of the moral economy of public water in Italy. In other words, this involves a water ethic, seen as a legislative and value frame of reference, deeply rooted in public opinion and, therefore, essential to define the legitimacy of any debate and policy concerning water resource management in the country. The reasoning and proposals surrounding "virtual water" must also address these principles and act consequently.

10 Virtual Water and Mobilisation in Italy

In Italy, the popular interest and mobilisation on water-related issues offer technicians and experts in water management a fertile ground to communicate and spread their knowledge and proposals, with the aim to guide decision-making in public policies, the strategies of the private sector and the demands of civil society, as suggested in the World Water Development Report 2009 (United Nations 2009). This is also valid for researchers and scientists involved in calculating the water footprint and in analysing virtual water flows, that is, the volume of water needed to produce the goods and services used by a country, by a city or by a single person, and the water resources which are transferred from one geographical area to another through the trade in these products.

The idea and the thinking underlying the concept of virtual water present several affinities with the moral economy of water as a commons which consolidated in Italy over the last years. First of all, theoretically, it seems to emerge a common criticism of the neoliberal paradigm and of the economic reductionism which it proposes. The focus on consumption and on lifestyles and ecological awareness are at the centre of the concepts of virtual water and water footprinting, and instead of calling for a balance in the economic parameters and management efficiency, they invite considerations on social equity and ecological sustainability. Secondly, the international awareness of the Italian water movement and the attention to global virtual water flows converge in the need to include the Italian water resource management debate in the context of global phenomena and processes. Finally, in practice, the research on both virtual water and the Italian water movement is aware of the need to ground future political project and initiatives on a thorough and innovative action aimed at the definition of new theoretical paradigms and popular understandings of water issues.

At the same time, because of its originality and relative novelty for the Italian public and activists, the concept of virtual water has the potential to provide new ideas for the Italian debate concerning water resource management, introducing new perspectives and posing questions and new challenges to institutions, movements and citizens which have participated to the mobilisation for water as a commons.

11 Strengthening the Ecological Dimension in the Italian Debate

In Italy, over the last years, the debate on water resource management has been oriented and framed mainly by legal and economic knowledge and notions. The discussions have therefore been focused on the "public/private", "state/market" and "commons/economically relevant services" dichotomies. Curiously enough, the technical and engineering paradigms, which have traditionally inspired the "hydraulic missions" of many civilizations in different parts of the planet, seem to have become secondary. The approach on which the idea of virtual water is based also allows other issues to be included in the debate, enhancing the contribution of other disciplines, such as geology, agronomy and commodity economics. The inclusion of new variables and the focus on other phenomena and processes make it easier to adopt an ecological and ecosystem perspective. The virtual water approach provides the opportunity to widen the field of research and action, overcoming the anthropocentric approach that has shaped the debate on water management in Italy. This, over the last years, has generally been restricted to water seen in terms of local services for civil and domestic uses. By unveiling the volumes and flows of water actually contained in our daily goods and services, the idea of virtual water allows us to create a link between the civil uses and other production sectors, especially in agriculture, between the policies of management of water services and the strategies for agricultural development, energy supply, and land use and conservation. This link is essential to encouraging the shift from an anthropocentric approach to an eco-centred one, where human uses and consumption are encompassed in the much broader context of the ecosystem, as suggested by the paradigm of "ecological water management" (Postel and Richter 2003).

12 Redefining Territorial Borders and "Local" Communities

Furthermore, the idea of water footprinting highlights the global interdependence determined by the virtual water flows implicitly embedded in goods and services traded internationally and consumed locally. The attention on the water cycle and production cycles raises the opportunity to critically rethink about territorial borders, as well as about the idea of belonging to so-called local communities. This awareness provides a means to overcome localisms and egoisms which risk slipping into limited or identity-centred interpretations of the notion of the commons and of the territory. Therefore, it appears that the concepts of virtual water and water footprinting require an approach that goes beyond the simplification of mere volumetric measurements, of which some production activities, exporting virtual water, would actually be the cause of scarcity in some countries of origin (Wichelns 2010). The shift from "local communities" to "water communities" is, instead, based on research that goes into a more in-depth analysis of the sustainability of the single production processes, with a view to sharing the risks (Pegram et al. 2009) among the different production sectors, subjects of the political community, and regions and populations belonging to different countries. From this point of view, any local management of water resources and services must work within a logic of involving all sectors and, first and foremost, the agricultural sector—the main consumer of water—and as well taking into account the ecosystems at the most suitable scale, which often overcome or cross-over administrative and political borders.

13 Strengthening and Enlarging the Network

Finally, the virtual water concept invites to reflect about the social and political alliances that promoted the Italian mobilisation for water as a commons, as well as about the subjects which can legitimately intervene in debates on water politics.

Above all, the thinking on virtual water has provided the opportunity to strengthen relations and involvement in the Italian mobilisation of subjects and realities particularly interested in the issue of water saving and lifestyles—local entities, environmental associations and consumers movements. In this perspective, there is undoubtedly the room for a more explicit and scientifically sound inclusion of virtual water notion and issues within, education, awareness campaign and training initiatives promoted by the Italian water movement. By deepening the analysis of the existing links between the development model, productive strategies and water resource management, the Italian water movements might broaden its scope and action, including for instance actors dealing with agriculture, food, land conservation and international trade—farmer associations, producers cooperatives, fair trade stores and the retail chains.

Finally, the focus on production processes which distinguishes the virtual water approach provides the Italian debate on water politics with probably the most controversial, but at the same time intriguing, input. On the one hand, the moral economy of water, which consolidated in Italy over the last years and which has been officially legitimised by the referendum success, has affirmed the nature of water as a commons, acknowledging public (the state, local entities) or civil (movements and citizen committees) players as the legitimate subjects in defining water policies and ensuring water management. On the other hand, the analysis of the impact on water resources of the production processes and distribution of goods and services promoted by the notion of virtual water directly calls into question the economic players, in most cases private. The FAO data show how, globally, the agricultural sector is the main consumer of water, using for irrigation purposes 90 % of total withdrawals, and how most of this consumption (80 %) is private (Allan 2013). By explaining and measuring the impact and the responsibility of private companies and commerce in terms of water consumption, the idea of virtual water offers these subjects the opportunity to draw up more efficient production strategies, economically and environmentally. The private operators that have opted for this solution, such as Barilla and Mutti described in this volume, have also gained the opportunity to be legitimately involved in the debate on water management in the eyes of the public.

Therefore, virtual water opens up an interesting space to involve private companies and economic actors producing services and goods in formulating policies and strategies for an integrated and ecological management of water. Not so much with the concealed purpose of letting back in through the backdoor the interests and ideas which the June 2011 referendum had closed out, but rather with the intention to include the actors that withdraw and consume the most water, in a framework of rights and duties concerning a consistent and shared sustainable resource management.

In conclusion, the remarkable public interest and the extraordinary citizen participation that featured the Italian mobilisation for water represent a potentially fertile and receptive ground for an awareness-raising campaign and concrete initiatives on the issue of virtual water. The idea of virtual water presents different contact points with the moral economy of water as a commons consolidated over the years within the Italian public opinion—the focus on the international and global implication of water management, the rejection of economic reductionism and the awareness of the importance of changes in cultural and lifestyle paradigms.

At the same time, the idea of virtual water invites a widening up of the horizons in the present national debate on water, moving in different directions. First, the idea of virtual water urges us to go beyond the anthropocentric approach focused on civil and domestic uses, in favour of an eco-centred approach, forcing us to address the question of how to guarantee the sustainability of water consumption for all the production sectors in a wider context of an ecosystem balance. Second, the flows of and trade in virtual water raise the need to critically review concepts such as regional and local communities and to rethink their borders. Finally, the focus on virtual water consumption opens up interesting possibilities to involve new subjects in the legitimate debate on water politics and, in particular, to redefine the roles and responsibilities of the private sector. The success of these three operations will also depend on the ability to sum up the virtual water message in a motto as happy and effective as the one of the Italian water movement—water as human right and commons—and to build an analogous wide and committed coalition to support it.

References

Allan J. A. (2013). Food-water security: Beyond water and the water sector. In B. Lankford., K. Bakker., M. Zeitoun & D. Conway (Eds.), *Water security: Principles, perspectives, practices*. London: Earthscan.

Bersani, M. (2011). *Come abbiamo vinto il referendum. Dalla battaglia per l'acqua pubblica alla democrazia dei beni comuni*. Rome: Edizioni Alegre.

Birdsall, N., & Nellis, J. (2002). Winners and losers: Assessing the distributional impact of privatization, center for global development working paper n. 6. Washington DC. http://www.sciencedirect.com/science/article/pii/S0305750X03001414.

Carrozza, C. (2011). Italian water services reform from 1994 to 2008: Decisional rounds and local modes of governance. *Water Policy, 13*(6), 751–768.

Carrozza, C. (2012). I referendum di giugno: Una vittoria a metà. In A. Bosco & D. McDonnell (Eds.), *Politica in Italia. I fatti dell'anno e le interpretazioni. Edizione 2012*, (pp. 257–274). Bologna: Il Mulino.

Diamanti, I. (2011, June 27). *Il movimento che rende visibile il cambiamento nel paese*. Rome: La Repubblica.

Fondazione Istituto Cattaneo (2011). *Il significato politico dei referendum*. www.cattaneo.org.

Faris, S. (2012, June 15). *Va in scena l'autogestione*. Italy: Internazionale.

Massarutto, A. (2010). La gestione dei servizi idrici privata … del buon senso. *L'amministratore locale, 1*.

Mattei, U., Reviglio, E., & Rodotà, S. (Eds.) (2007). *Invertire la rotta: Idee per una riforma della proprietà pubblica*. Bologna: Il Mulino.

Mattei, U., Reviglio, E., & Rodotà, S. (a cura di). (2010). *I beni pubblici. Dal governo democratico dell'economia alla riforma del codice civile*. Rome: Accademia Nazionale dei Lincei.

Molinari, E., & Jampaglia, C. (2010). *Salvare l'acqua. Contro la privatizzazione dell'acqua in Italia* (p. 210). Milan: Feltrinelli.

Pegram, G., Orr, S., & Williams, C. (2009). *Investigating shared risk in water: Corporate engagement with the public policy process*. WWF-UK: Godalming.

Petrella, R. (2001). *Il Manifesto dell'acqua*. Turin: Edizioni Gruppo Abele.

Petrini, C. (2010, May 6). *Acqua. Quando il bene comune diventa una merce*. Rome: La Repubblica.

Petrini, C. (2008, Nov 14). *La soluzione migliore è lasciarla alle città*. Rome: La Repubblica.

Postel, S., & Richter, B. (2003). *Rivers for life: Managing water for people and nature*. Washington DC: Island Press.

Rodotà, S. (2012, Jan 5). *Il valore dei beni comuni*. Rome: La Repubblica.

Thompson, E. P. (2009). *L'economia morale delle classi popolari inglesi nel secolo XVIII*, et al. Edizioni, Milan (ed. or. 1971).

United Nations, World Water Assessment Programme. (2009). *WWDR3: Water in a changing world*. London: Parsi, UNESCO.

Wichelns, D. (2010). Virtual water: A helpful perspective, but not a sufficient policy criterion. *Water Resources Management, 24*(10), 2203–2219.

Part III
The Italian Case

Water Resources in Italy: The Present Situation and Future Trends

Monia Santini and Maria Cristina Rulli

Considering the assessment on water resources shown in Table 1 and a consumption of 92 m^3 per capita annually for the period 1996–2007 (more than the 85 m^3 average in the EU27 countries), Italy seems to be highly vulnerable to any reduction in its water availability.

Besides the temporal variability in water resource availability/demand, there are also differences at sub-national spatial level. The interaction between the climatic, topographic and geological characteristics and human activities makes the Italian context highly heterogeneous. While the northern regions, despite increased water exploitation due to their predominantly intensive farming and industrial concentration, can rely on plentiful and regularly available water resources, in the south this availability is approximately half of the need because of a combination of low rainfall and high temperatures which increase hydrological losses due to evapotranspiration. Apulia, Sicily and Sardinia receive 40–50 % less rainfall than the wetter regions and just covering 10–20 % of their water needs.

There are different factors that influence both current and future water resource availability:

- high and growing consumption due to the combination of an increase in demand (for example due to population growth, estimated at about 4 million over the last 30 years) and a decrease in natural recharge caused by natural and human factors;
- pollution of water reserves;

M. Santini (✉)
Division Impacts on Agriculture, Forests and Ecosystem Services,
Centro Euro-Mediterraneo sui Cambiamenti Climatici, Viterbo, Italy
e-mail: monia.santini@cmcc.it
URL: http://www.cmcc.it

M.C. Rulli
Department of Civil and Environmental Engineering, Polytechnic of Milan, Milan, Italy

© Springer International Publishing Switzerland 2015

M. Antonelli and F. Greco (eds.), *The Water We Eat*,
Springer Water, DOI 10.1007/978-3-319-16393-2_11

Table 1 Water resource data in Italy

Area	30.134	km^2
Population	60,789,000	inhabitants
Average annual rainfall	832	mm/year
Internal renewable water resources	182.5	km^3/year
External renewable water resources (Switzerland, France, Slovenia)	8.8	km^3/year
Total renewable water resources	191.3	km^3/year
Potential water availability per capita	3,147	m^3/year
Water resources effectively available and useable	123.0	km^3/year
Water loss distribution	36.2	%
North	33.7	%
Centre	39.1	%
South and Islands	55.0	%
Water resources effectively used	45.4	km^3/year
Water resources effectively consumed per capita	747	m^3/year
Water use*		
Agriculture	44–60	%
Industry	25–36	%
Domestic use	15–20	%
National withdrawals for drinkable water use	9–11	km^3/year
Drinkable water origin		
Groundwater	85.6	%
Surface water	14.4	%

Source: AQUASTAT, March 2015; ISTAT, 2011; CONVIRI, 2011.
* The range comes from a data analysis: AQUASTAT; Conferenza Nazionale delle Acque (1972) and further updating by the Italian Ministry for Agriculture and Forestry (1990); ISTAT (1991); IRSA-CNR (1999); Legambiente (2007); Italian Ministry for Agriculture and Forestry (2004).

- changes in types of consumption (Italy leads by far in the consumption per capita of mineral water, accounting for 70 % of the total);
- the weakness in the system of water distribution, recycling and reuse (only 0.2 % of water actually available comes from desalination plants).

Despite the inter-annual variability in water resource availability over the 30-year period 1971–2000, a general trend can be clearly identified in terms of decreasing rainfall and effective infiltration—that is, the percentage of rainfall percolating through the soil and potentially feeding the underground water resources (Fig. 1).

The analysis of the climate projections, based on short- and long-term simulations, reveals a worsening in water resource availability affecting agriculture in the short term, and the groundwater recharge in the long term.

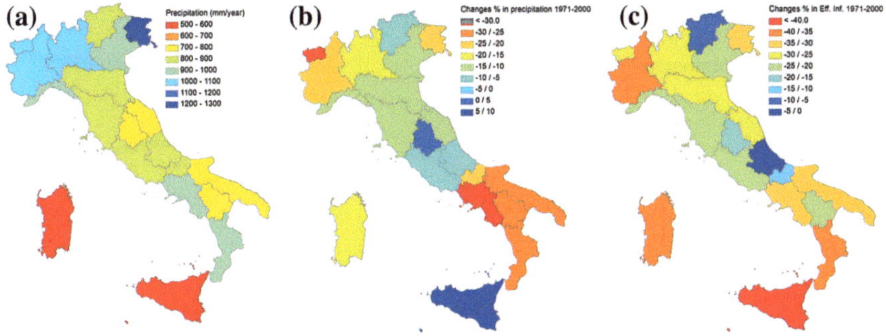

Fig. 1 Maps of the annual average rainfall (**a**), percentage variation in rainfall (**b**) and percentage variation in effective infiltration (**c**) along the period 1971–2000 for the different Italian regions. *Source* Venezian Scarascia et al. (2006)

Situated at the centre of the Mediterranean basin, Italy is particularly susceptible to climate change presenting climate shift towards warming (following the global trend) and progressive drying and desertification (Giorgi 2006).

It should be remembered that, where climate change is concerned, there is a widespread scientific debate on the reliability and uncertainty of climatic projections for a given region. This is due to the fact that many quite different models exist (GCMs—global circulation models) to simulate coupled atmosphere–ocean dynamics even under different greenhouse gas emission forcing based on alternative IPCC development scenarios. In the analyses presented by Giorgi (2006), the uncertainty related to future climatic trends is quite reduced thanks to the probabilistic analysis relying on an ensemble of simulations conducted by combining 20 different GCMs and 3 different greenhouse gas emission scenarios. Indeed, the agreement among the different ensemble simulations was evaluated and the likelihood of climate scenarios was assessed. Therefore, the resulting maps from simulation ensembles show the findings in terms of consensus among the different members in reproducing a decreasing rainfall trend (García-Ruis et al. 2011; Giorgi and Lionello 2008) and, consequently, a decreasing surface run-off, more marked in the southern Mediterranean regions and concentrated in the summer period.

Climate simulations carried out by the Centro Euro-Mediterraneo sui Cambiamenti Climatici (CMCC) over Italy with high spatial resolution models confirm that the summer season will be more likely suffering from increased temperatures and reduced rainfall (Comegna et al. 2012), with obvious consequences for the hydrological cycle.

Added to these climatic trends, it is expected a probable rise in water demand due to increases in population and production use, particularly for agriculture. The ISTAT projections for 2065 (http://www.istat.it/it/archivio/48875/), taking into account plausible trends in birth/mortality rates and immigration (compared to 2010), foresee a demographic growth of almost one million to about 9 million people, respectively, for two of the three scenarios considered (Fig. 2).

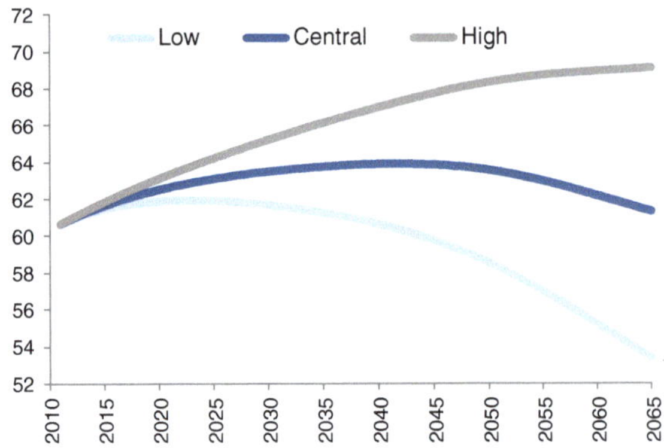

Fig. 2 Resident population in Italy (mill.) according to different ISTAT demographic scenarios

Moreover, Flörke and Alcamo (2004) estimated an increase in irrigated areas of up to 27 % from 2000 to 2030 in the base scenario case, if assuming a continuation of the same current environmental conservation (including water resource) policies.

An ensemble of 24 simulations carried out by the CMCC using a dynamical model on potential vegetation (LPJ; Sitch et al. 2003), which also takes into account the water balance in the simulations, and combining different model configurations and parameterisations with alternative climate projections up to 2050 (all under the IPCC-AIB scenario, with a very rapid global economic growth, a global population reaching a peak mid-century and then decreasing, a rapid introduction of new and more efficient technologies and a balanced distribution among the different energy sources), led to analysing the likelihood of the expected variation in the water balance components like run-off, evapotranspiration and soil moisture, considering the statistical distribution of the simulation results (e.g. Santini et al. 2014). A high spatial variability in the hydrological cycle components was revealed, reinforcing the need to differentiate the future water resource management choices for the various regions, considering both their water availability and degree of vulnerability. Similar spatiotemporal analyses, obtained by combining sophisticated models simulating climate and hydrological dynamics with projections of socio-economic and land-use trends, contributed to an increase of scientifically sound information about the likely effects of future scenarios on water resource allocation among the different uses (agriculture, industrial, civil) and on the feedbacks between this allocation and the hydrological cycle. In particular, in such a diversified and vulnerable country as Italy, this type of analysis, when able to differentiate water availability, deficit, demand and supply at national to sub-national level, and in the short to long term, could be very useful for stakeholders and policy-makers in identifying and prioritising regulations, initiatives, and/or investments towards adaptation to climate and socio-economic changes and/or mitigation of their impacts.

References

Conferenza Nazionale delle Acque (1972). *I problemi delle acque in Italia*, a cura del Senato della Repubblica, Roma.

Comegna, L., et al. (2012). Potential effects of incoming climate changes on the behaviour of slow active landslides in clay. *Landslides*. doi:10.1007/s10346-012-0339-3.

CONVIRI—*Commissione nazionale per la vigilanza sulle risorse idriche, Rapporto sullo stato dei servizi idrici*, Rome, December 2011.

Flörke, M., & Alcamo, J. (2004). *European Outlook on Water Use*, Centre for Environmental Systems Research—University of Kassel, Kassel, Final Report, EEA/RNC/03/007.

García-Ruiz, J. M., et al. (2011). Mediterranean water resources in a global change scenario. *Earth-Science Reviews, 105*(3–4), 121–139.

Giorgi, F. (2006). Climate change hot-spots. *Geophysical Research Letters, 33*(8), L08707. doi:10.1029/2006GL025734.

Giorgi, F., & Lionello, P. (2008). Climate change projections for the Mediterranean region. *Global and Planetary Change, 63*, 90–104.

IRSA-CNR (1999). *Un futuro per l'acqua in Italia*, Quaderni di ricerca IRSA-CNR, n. 109, Rome.

Italian National Institute of Statistics (ISTAT) (1991). *Approvvigionamento idrico, fognature e impianti di depurazione in Italia: anno 1987*. ISTAT, Roma.

Italian National Institute of Statistics (ISTAT) (2011). *Giornata Mondiale dell'acqua. Le statistiche dell'ISTAT*. Roma.

Italian Ministry for Agriculture and Forestry (1990). *I problemi delle acque in Italia. Aggiornamento al 1989 dei risultati della Conferenza Nazionale delle Acque*. Bologna, Edizione Agricole.

Italian Ministry for Agriculture and Forestry (2004). *Irrigazione sostenibile, la buona pratica irrigua*, L'informatore agrario, pp. 316.

Legambiente. (2007). *L'emergenza idrica in Italia. Il libro bianco di Legambiente*, Dossier May.

Sanchez-Gomez, E., Somot, S., & Mariotti, A. (2009). Future changes in the Mediterranean water budget projected by an ensemble of regional climate models, *Geophysical Research Letters, 36* (21). doi:10.1029/2009GL040120.

Santini, M., Collalti, A., & Valentini, R. (2014). Climate change impacts on vegetation and water cycle in the Euro-Mediterranean region, studied by a likelihood approach. *Regional Environmental Change, 14*(4), 1405–1418.

Scarascia, M. E., Di Battista, F., & Salvati, L. (2006). Water resources in Italy: Availability and agricultural uses. *Irrigation and Drainage, 55*(2), 115–127. doi:10.1002/ird.222.

Sitch, S., et al. (2003). Evaluation of ecosystem dynamics, plant geography and terrestrial carbon cycling in the LPJ dynamic vegetation model. *Global Change Biology, 9*(2), 161–185.

Todorovic, M., Caliandro, A., & Albrizio, R. (2007). Irrigated agriculture and water use efficiency in Italy. *Options Méditerranéennes, Serie B: Studies and Research, 57*, 101–136.

The Globalisation of Food and Water: The Italian Case

"Water in Food" Group, Politecnico di Torino

Paola Allamano, Pierluigi Claps, Paolo D'Odorico, Francesco Laio, Luca Ridolfi and Stefania Tamea

Food security, intended as the agricultural production capacity to satisfy the nutritional needs of the world's population, is closely tied to water availability, the latter being essential for the production of any type of food. The water volume required to produce an established quantity of food is called the "virtual water content" and is the amount of water virtually embedded in the good, though not physically present in it. Italy is an exemplary case of high virtual water consumption and dependence on food imports. It is among the leading countries in the world for net virtual water importation, with a high per capita consumption and a persistent reduction in land surface used for agricultural production. Local contradictions arising from the global food supply model are the issue dealt with in this contribution, with particular reference to Italy. Agriculture, often neglected in press campaigns on water saving, is, instead, the main consumer of water; the production of food goods corresponds to a global consumption of about 80–90 % of the total water used to satisfy anthropic demand (Falkenmark and Rockström 2004). The water volume required to produce an established quantity of food is called the "virtual water content" and is the amount of water virtually embedded in the good, though not physically present in it (Allan 1993). For example, it is calculated (Hoekstra and Chapagain 2008) that 1600 l of water are required to produce one kilo of bread and 15,400 l for 1 kg of beef. An individual consumes, on average, 2000 l of water a day, about one thousand times the daily per capita need for drinking water.

Apart from some rare exceptions, in past traditional economies, most food was produced and consumed locally and, consequently, there were no great transfers in virtual water. In these systems, the population growth in a given geographic region was limited by the local water resource availability, tempered by the level of efficiency reached in managing this resource. In recent times, however, the global market trade in food has allowed local populations to be free from the restrictions of local water resource availability, allowing some populations to exceed the food

P. Allamano (✉) · P. Claps · P. D'Odorico · F. Laio · L. Ridolfi · S. Tamea
Research Group "Water in Food", Department of Environmental, Land and Infrastructures, Engineering, Politecnico di Torino, Turin, Italy
e-mail: Paola.allamano@polito.it

© Springer International Publishing Switzerland 2015

M. Antonelli and F. Greco (eds.), *The Water We Eat*,
Springer Water, DOI 10.1007/978-3-319-16393-2_12

need limitations imposed by the water available in loco (Allan 1998). International trade results in the virtual transfer of water from food production areas to importing regions, resulting in a disconnection between demographic expansion and locally available natural resources (Carr et al. 2012).

Scientific research and debates on policies regarding the virtual water trade have shifted towards the idea that, by importing virtual water, countries relatively scarce in water could preserve or use their water reserves quite efficiently (Hoekstra and Chapagain 2008). Not only does international trade lead to a more efficient allocation of goods production with a high water content, but it also contributes to decreasing the use of water at a global level. Thanks to the production specialisation imposed by trade, a growing part of high water content goods can be produced in countries with a more efficient water use and exported to less efficient countries. Indeed, according to most studies on the issue, the net effect of the world virtual water trade appears to reduce the overall use of water (De Fraiture et al. 2004; Oki and Kanae 2004; Hoekstra and Chapagain 2008).

However, in certain cases, the empirical evidence clashes with this theory. The studies on virtual water flows highlighted, for example, that major trade occur from relatively water-short regions to regions with abundant water resources. Well-known examples include China (Hoekstra and Chapagain 2008) and India (Verma et al. 2009), but also Africa emerges as an exporter of virtual water to Italy, despite the limited water availability of many African countries. This can be explained by the fact that the cost of water represents only a small part of the production cost, together with the fact that water is not generally traded at market prices (Hoekstra and Chapagain 2008). This situation derives from various "externalities", that is situations where the value of the traded goods does not take into account all the actual costs involved in their production, and in these circumstances, a water resource use results in being incompatible with its availability. The absence of a suitable and common system of rules to define water resource economic value leads to difficulties in assessing the possible over-exploitation of the resource in fragile contexts from socio-economic and environmental perspectives. This phenomenon tends being exacerbated by world demographic pressures, which result in an increasing water resource demand, with water becoming a key factor for food security and community well-being (Rosegrant et al. 2002; World Economic Forum 2011); Hoekstra and Mekonnen 2012). Scientists, politicians, decision-makers are increasingly realising that the development of the right strategies to meet ecosystem water requirements on the one hand, and world's population demand on the other, is one of the main environmental challenges of this millennium (Falkenmark and Rockstrom 2006; Hanjra and Qureshi 2010; Vörösmarty et al. 2010). Within this framework, it is crucial to understand the globalisation of water (Hoekstra and Chapagain 2008) induced by the trade in virtual water. Furthermore, even though the problem is intrinsically global, due to the increase in the trade in food goods, there can be very different implications for individual countries. Italy is an exemplary case of high virtual water consumption and dependence on food imports. It is among the leading countries in the world for net virtual water importation, with a high per capita consumption and a persistent reduction in land surface used for agricultural production. Local contradictions arising from the global food supply model are the issue dealt with in this contribution, with particular reference to Italy.

1 General Context

The final objective of the analyses which we will present in the following is to reconstruct, on a single country level or global level, the terms of the virtual water balance, that is the total virtual water volume involved in production, consumption, importation and exportation. Three types of data are required to obtain the virtual water volumes. The first concerns the world trade in goods where the data are updated every year for hundreds of products and refer to each country's imports and exports. The second type, instead, refers to the internal production of the same goods (also here the quantity produced in each country and for each year is available). The third type of data pertains with the virtual water volume contained in one unit of each good (for example, a tonne of wheat or milk) for each considered country. This detail is necessary as the virtual water content depends on many factors, such as climate, soil features, agricultural methods used and the irrigation infrastructures in the production regions. Moreover, it is necessary to know both the origin of the water used in producing the goods, so as to differentiate the so-called green water (rainwater) and blue water (irrigation water), and the pollutant load involved in producing the goods to estimate the grey water (the water required to maintain the concentration of fertilisers and pesticides under the legal thresholds). Therefore, it is easy to understand that assessing the water content of a good is a somewhat delicate operation. Once these three types of information are known, each good's weight can be converted, be it produced, exported or imported, into its corresponding water content, and then, the global virtual water flows and the relative balances for each country can be reconstructed, from which the corresponding water footprint can be obtained (Hoekstra and Mekonnen 2012). The most used databank (which we will refer to) is managed and made available by FAO (UN Food and Agricultural Organisation, www.faostat.fao.org). Specifically, it provides information on 309 agricultural products surveyed over the period 1986–2010. This period, apart from cases where literature data are reported, will be the one considered in the following analyses.

In studying the recent history of virtual water trade, some figures will help us in providing an immediate picture of the phenomenon. The temporal evolution of the traded volumes among all the countries of the world (Fig. 1) highlights both the large volumes of water traded, thousands of cubic kilometres, and their marked increase over the last decades. To appreciate the actual volume involved, it must be taken into account that, in 2008, it corresponded to about 50 times the water volume of the Po River annual flow at the mouth. The increase in time is also extremely relevant, as seen by the volume doubling over a span of 23 years. Since in the same period the world population did not increase at the same rate, it can be concluded that the

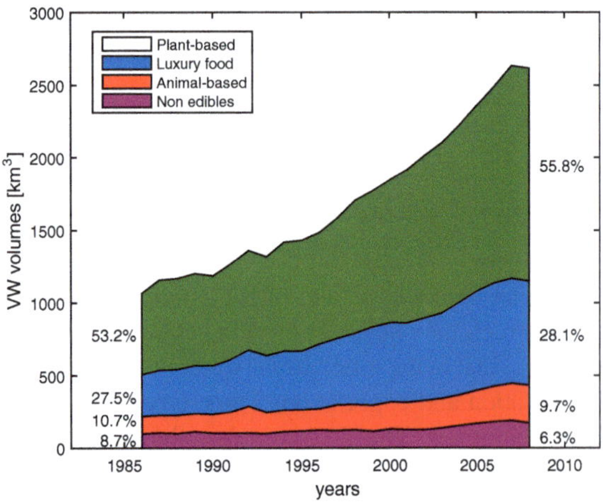

Fig. 1 Temporal evolution of virtual water volumes exchanged worldwide, subdivided into the four main categories. *Source* Carr et al. (2013)

average volumes per capita of traded virtual water clearly increased from about 210 to 320 m^3/year. The same figure also shows the contribution of different product macro-categories, from which it is seen that more than a half of virtual water is traded through basic food plant products (such as cereals), a little more than 28 % regards agricultural goods non-essential for nutrition (e.g. coffee, cocoa, etc.), while a little less than 10 % regards meat and animal-origin products. Concerning the latter, it can be noted that even if the different percentages remained substantially unchanged over the years, the percentage regarding meat tended to slightly decrease.

A recent study conducted by Carr et al. (2013) has shown how the geography of net virtual water flows has changed over time (1986–2006). The basic data emerging from the study is that there are only a few main exporters that have substantially remained unchanged (Canada, USA, Australia, Argentina, Brazil, Indonesia, etc.), with the appearance of Ukraine during the last decade, while most countries are net importers—among these the Mediterranean and European countries emerge. The larger surface areas of some net exporter countries (e.g. Canada and Australia) do not betray the fact that only a minority of the population holds positive net flows in exportation. This involves about 6–8 % of the world population with this percentage having remained more or less constant in time. Moreover, by clear demographic weight, China especially deserves to be noted as it has increasingly become a virtual water importer; also India is worth being mentioned, where importation/exportation has tended to become balanced, however, at the cost of a marked over-exploitation of "blue water" reserves (groundwater).

Observing that each good traded between two countries corresponds to a virtual water flow, it is easy to imagine how virtual water trade gives rise to a very complicated global network. The evidence is in the large number of connections, about 15,700, compared to the number of hubs (i.e. the countries), about 200. It is interesting to note that, also from this perspective, there has been a marked increase in globalisation. Carr et al. (2013) showed that the degree of connections among countries has markedly increased over time and that countries initially quite marginalised, particularly in Africa, have begun to be more connected. The increase in the number of connections should not convey the image of a network increase due to accumulation, that is a network where each time new links are added without changing the pre-existing ones. On the contrary, the network displays a high plasticity; that is, many of the existing connections disappear, while others emerge again.

The information reported above unequivocally demonstrates that international virtual water trade is becoming increasingly important in the water balance of single countries, making water resource management an issue which is becoming less local and more global.

2 The Italian Case

Italy is a prototypical case of high virtual water consumption and heavy dependence on food imports, being one of the main countries in the world for net virtual water importation. Therefore, it is interesting to analyse the country's balance of virtual water flows and their composition regarding the agri-food product categories generating them. Moreover, it will be possible to outline the geography of the virtual water flows and the globalisation of water, describing the international trade networks which have their origin or destination in Italy.

2.1 The Virtual Water Balance

In 2010, Italy traded almost 130 km^3 of virtual water abroad, importing 91.4 km^3 and exporting 36.8 km^3, with local agricultural production use of 65.9 km^3. The ratio between the volumes traded and those produced in loco is, therefore, slightly less than two, evidence of an intensive commercial activity associated with a less important productive activity. And this is not a peculiarity for year 2010 only. The ratio between volumes traded and produced in loco has in fact increased over the last 25 years. A more complete picture of the situation is found in Fig. 2, which shows the annual volumes of virtual water imported to and exported from Italy in the period 1986–2010, together with the amount consumed by the population and that used for the internal production of agri-food goods.

It can be seen that the volumes of virtual water traded abroad have significantly increased over the last decades. The increase in exports has been continuous and steady over time, and the volume traded tripled in the 25 years under study. Imports also markedly increased but with some fluctuations in the more recent years, with a total increase of 80 %. During the period considered, Italy greatly increased its dependence on the international market, to the extent that imports exceeded the volume used for internal production, almost reaching the virtual water volume consumed by the Italian population.

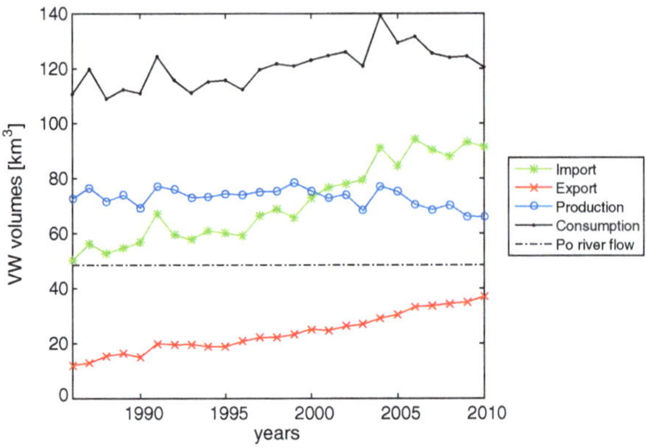

Fig. 2 Virtual water balance of Italy in the period 1986–2010, and comparison with the total annual flow of the River Po at the mouth. *Source* The authors

The virtual water used in agri-food production remained almost constant during the period, with only a slight reduction. This trend is due to three combined factors: the cultivated area in Italy has reduced by more than 20 % in the period 1986–2010 (from 127,000 to 97,000 ha, according to FAOSTAT data); an increase in crop yield (agricultural production per unit of area) occurred along witha shift to crops requiring more water. The outcome of these three factors results, as mentioned, in a slight decrease in the water volumes used in agriculture. Still referring to Fig. 2, it is interesting to note that the virtual water relating to internal consumption, obtained from the difference between net importation (import minus export) and the use in production, in the 25 years studied, shows a growing trend, to a large extent due to the increase in the Italian population.

Comparing the virtual water volume with the total annual flow of the largest Italian river, the Po River (1540 m^3/s or 48.6 km^3/year), it is observed that, in 2010, Italy used in food production a water volume which is about 1.5 times the annual volume flowing from the Po into the Adriatic Sea, while, in 2010, Italian virtual water importation was almost double the total annual flow of the Po. To integrate and better understand the comparison of virtual water flows with the Po River, it should be taken into account that the Po's basin covers almost a quarter of the Italian surface and that 48.6 km^3/year of water corresponds to almost half the total flow theoretically available in all Italian rivers, estimated at about 104 km^3/year. This total was estimated using a simplified model which, through a global inflow/outflow ratio, provides an estimate of the total surface water resources available for the Italian area. The model is based on a multiple regression analysis using rainfall and the Budyko Index (the ratio between potential evapotranspiration and the average annual precipitation) as the runoff explicative variables and is based on about 300 catchment basins considered under the CUBIST project (project funded by the Ministry for Education, Universities and Research, www.cubist.polito.it) which cover almost all Italian land surface. By applying the regression analysis model for all of Italy, it is estimated that the average annual runoff is about 352 mm (that is, 104 km^3), for an average annual precipitation of about 848 mm (that is, 252 km^3). Based on these estimates, the approximate 90 km^3 of imported water appears to be an extremely high figure, which raises serious questions about the long-term sustainability of Italian water consumption.

2.2 Analysis Per Category

Up to now we have considered the virtual water flows independently from their associated agri-food goods. However, interesting aspects emerge if we take into account the product typologies which are involved in virtual water trade. For this reason, four categories were individuated, including those products exported from Italy to other countries. The categories are (see Carr et al. 2013) as follows: edible plant-origin goods, animal-origin goods, luxury goods (such as coffee, cocoa, spices, etc.) and non-edible commodities (such as natural fibres). The main findings of this analysis can be seen in Fig. 3, which shows the temporal evolution of virtual water volumes for each category for the period 1986–2010.

Some general characteristics are evident: firstly, the plant-origin products represent the main goods for virtual water imports for all the period under study, up to 50 % of the total imported volume in 2010. Instead, the imports of animal-origin products

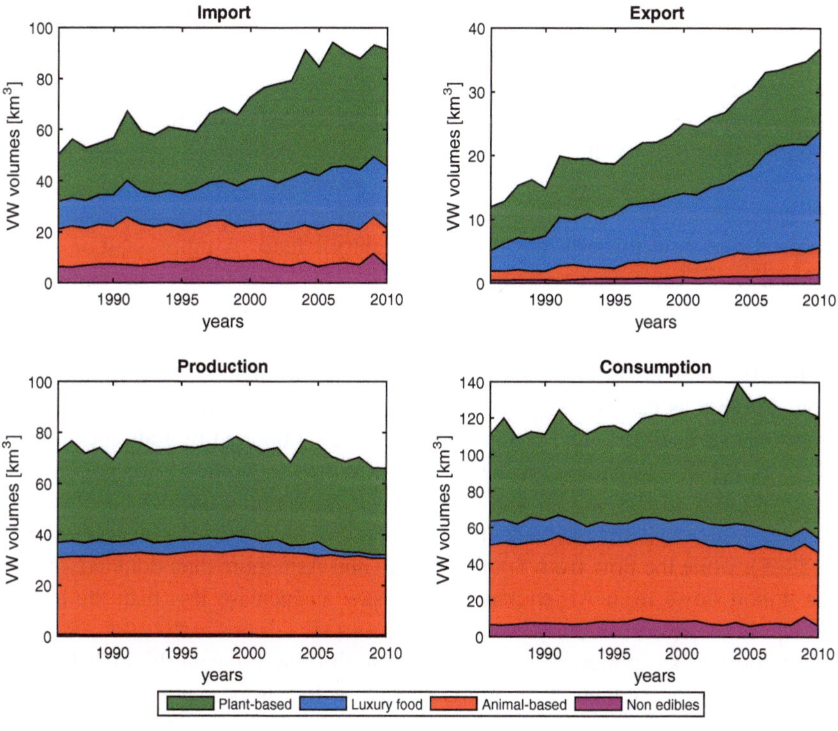

Fig. 3 Composition per category of the terms of the virtual water balance of Italy in the period 1986–2010. *Source* The authors

remained stable, despite the marked increase in total imports, which led to a reduction in the animal-origin importation percentage, possibly due to changes in commercial and eating trends. In the last years, the second main contribution to virtual water importation was linked to luxury goods, which increased significantly—from 21 to 26 % of total imports.

As for exportation, there are some basic differences compared to importation. The plant products dominated virtual water exports at the beginning of the period, whereas the luxury products took the lead after year 2000, reaching about 50 % of exports. The behaviour of this category evidences the importance offood processing industry in Italy (e.g. wine, coffee, pasta and baked goods). The animal or non-food products are a minor part of Italian virtual water exports compared to plant and luxury products and show a weak growth trend.

Finally, the use of virtual water in internal production mainly concerns plant- and animal-origin goods, which maintained quite constant volumes, while luxury goods underwent a recent decline. The use of water for non-food production is quite modest, with a consumption of 7–8 km^3/year which is satisfied by imports. A significant increase in virtual water linked to plant goods can be noted, from 47 to 66 km^3, while the other categories remained steady.

2.3 The International Trade Network

A geographic representation of virtual water flows which originate and end in Italy can be illustrated in a map showing the volumes traded with other countries. The volumes imported by Italy per year from each country are represented as river branches showing the country of origin from where the goods were imported. All the branches are then linked up in a hydrographic network grouping them into a single direct flow towards the importer country—Italy. A representation of this "virtual river" can be seen in Fig. 4 for 2010, where line thicknesses and colours indicate the ranking of the flow sizes. For a detailed assessment of the virtual water flows, Table 1 shows the imported and exported volumes in 1986 and 2010.

In 2010, Italy imported 91.4 km^3 of virtual water from all continents, but mainly from Europe, as can be seen by the thickness of the line grouping the European contributions in Fig. 4. The total virtual water flow imported by Italy increased by 82 % from 1986 (when it was only 50.3 km^3); however, the trend was not the same in all regions of the world. For example, North America (which also includes Central America) is the only region to have decreased its virtual water contribution (−28 %), while the flow from South America and Asia more than doubled. Instead, the import flows from Africa and Oceania saw an increase less than the average growth in the reference period.

As for Italy's exportation of virtual water to the rest of the world, in 2010 it was 36.8 km^3, with more than 70 % of this direct flow to European countries. The trend in exported virtual water was still more significant than for imports—the total flow in 1986 was a third of the 2010 flow. Also in this case, the variations of the different

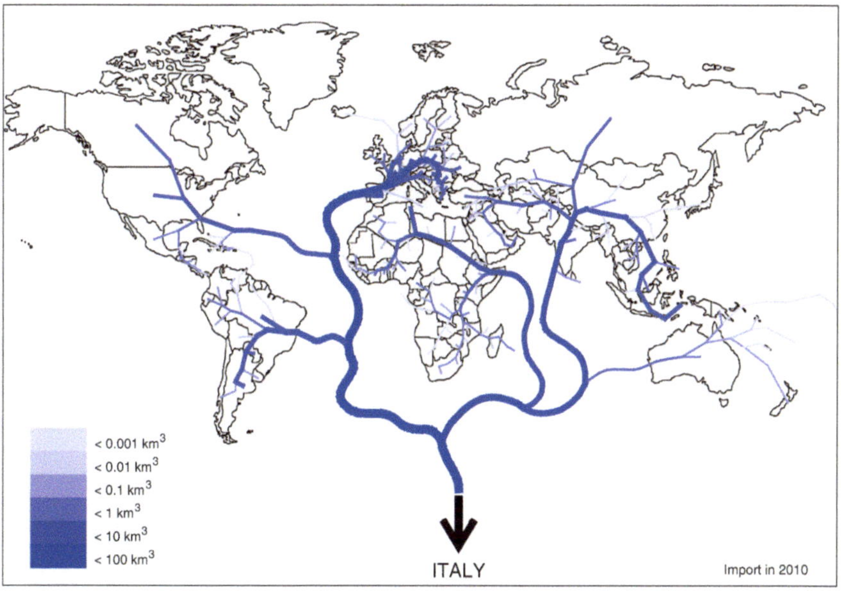

Fig. 4 The virtual water river flow towards Italy in 2010. *Source* Tamea et al. (2012)

Table 1 Origin and destination of Italian virtual water import and export in 1986 and 2010 (km^3/year)

	Import				Export			
	1986		2010		1986		2010	
Europe	28.2	56 %	54.6	60 %	6.7	56 %	26.5	72 %
Asia	4.5	9 %	11.7	13 %	1.2	10 %	5	13 %
Africa	5.9	12 %	8.2	9 %	2.7	23 %	1.5	4 %
North America	6.6	13 %	4.8	5 %	0.9	7 %	3	8 %
South America	4.5	9 %	11.4	12 %	0.4	4 %	0.3	0.90 %
Oceania	0.58	1.20 %	0.87	1.00 %	0.06	0.50 %	0.55	1.50 %
Total	50.3	100 %	91.4	100 %	12	100 %	36.8	100 %

Percentages are calculated based on the total in the last line
Source Tamea et al. (2012)

regions of the world were not the same—Africa and South America reduced their virtual water imports (−45 and −24 %, respectively), while the other continents more than tripled their flows. In particular, exports to European countries quadrupled, from 6.7 to 26.5 km^3. This increase was a consequence of the enlarging and strengthening of the Common European Market during the period studied.

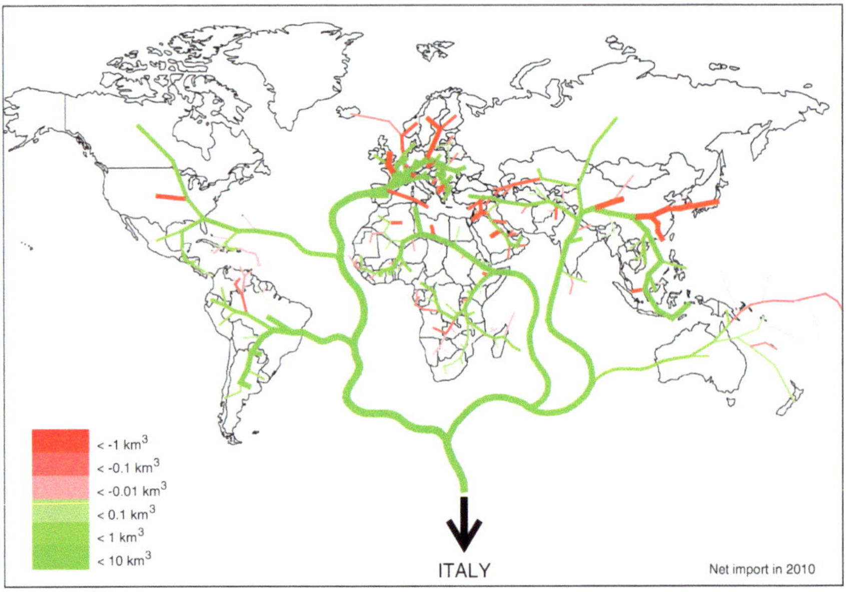

Fig. 5 Italian net import (import–export) of virtual water in 2010. *Green lines* indicate a net import and *red lines* indicate a negative net import (thus a net export) of Italy. *Source* The authors

The net import flow (import minus export) can be estimated by combining the imported and exported volumes from Italy in relation to each country for one year. Analogously to the geographic diagram in Fig. 4, the net virtual river can be constructed, where the positive or negative sign of the flows is identified by colour (Fig. 5). One can see how the net flow from all the continents is positive; that is, Italian virtual water imports exceed its exports, but there are also important negative flows (i.e. exports dominate) towards single countries such as the USA, UK, Northern European countries, China and Japan, where the line thicknesses show a very high volume of exported virtual water.

In the years between 1986 and 2010, there was an evolution in the trends of imports and exports. The network underwent only small changes, with a slight increase in the number of supplier countries and a more conspicuous increase in countries receiving exports. Some modifications occurred due to political/administrative changes in countries such as the ex-Soviet Union and ex-Yugoslavia, while other links, for example with the remote islands of the Pacific Ocean, were intermittent. The most significant variations compared to 1986 include the marked decline in net imports by (or the increase in Italian export towards) the USA, China, Japan, Oceania, the countries of southern African and Europe (even if the net European flows increased during the time, that is imports increased more than exports). Figure 6 shows the import and net import flows only concerning the European and Mediterranean area, in order to highlight the contribution of countries

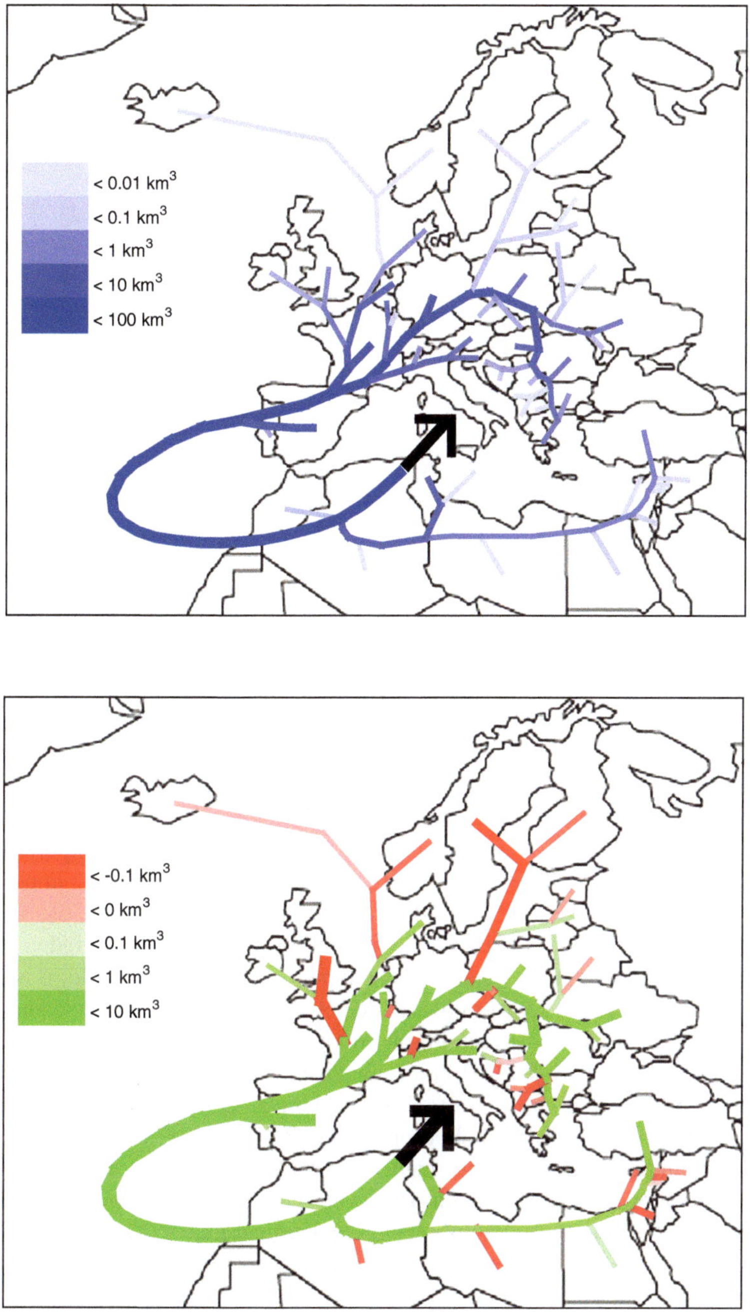

Fig. 6 Contribution of European and Mediterranean countries to Italian imports (*above*) and net imports (*below*) of virtual water in 2010. *Source* The authors

Table 2 Virtual water import (towards Italy) and export (from Italy) in 1986 and 2010 of some countries (in km^3/year)

	Import 1986		Import 2010		Export 1986		Export 2010	
Brazil	2.37	*# 5*	4.37	*# 4*	0.4	*# 8*	0.13	*# 39*
China	0.47	*# 26*	0.21	*# 49*	0.05	*# 32*	0.8	*# 13*
France	8.94	*# 1*	10.82	*# 1*	1.96	*# 1*	5.23	*# 2*
Germany	3.7	*# 4*	9.02	*# 3*	1.63	*# 2*	5.44	*# 1*
United States	4.89	*# 2*	1.98	*# 12*	0.74	*# 4*	2.19	*# 3*

In italics is the position of each country in the ranking for volumes traded with Italy
Source Tamea et al. (2012)

geographically closer to Italy. It can be seen that most of the virtual water imports within Europe originate from France, Spain, Germany and the net export flows concern countries such as the UK and the Nordic countries. In the Mediterranean area, Turkey and especially Tunisia move huge volumes of virtual water towards Italy, comparable to the main European countries, and explainable by the intense commercial trade and by the high water content of the traded products.

If the trade flows from Italy to some countries of particular interest are considered (Table 2), France remains the preferred partner for Italian imports, even if the increase in the import flow over the period is weaker compared to the world average (21 % compared to the average of 82 %). The USA has been reducing their contribution to Italian imports while, on the contrary, Brazil is taking on a leading role, with high growth rates over the last years. When exports are considered, Germany and France emerge as the main partners, with Germany in the last years tripling its contribution. Italy strongly increased its virtual water exports to the USA, with a total flow (1.98 km^3 in 2010) exceeding the corresponding import flow (see Fig. 5). Italy also significantly increased its penetration into the Chinese market, while flows towards Brazil significantly decreased in the period.

Finally, to better compare Italy and the other countries of the world, it is useful to consider the net virtual water flows traded by all countries during 1986 and 2010. Ranking the countries by decreasing net imports, it emerges that Italy was in 3rd position in 1986 and 5th in 2010, confirming the fact that Italy was among the main world net importers of virtual water. If the ranking is repeated based on the net per capita flow, it is noticed that Italy drops to 48–49th position, as in this case some smaller countries tend to register higher net per capita imports, thus making the interpretation of the results difficult. However, by totalling the population living in countries with a net per capita importation higher than Italy's (47 and 48 countries in the two cases), we obtain 1 and 3 %, respectively, of the world population. This shows that most of the world's population has a net per capita virtual water importation lower than the Italian average.

3 Conclusions

The concept of virtual water is a new paradigm for the understanding of the complex dynamics relevant to the use of water resources by humans. In particular, the use of this indicator results in highlighting the main role agriculture plays in the resource's consumption and, as well, in recognising how water resource management has now become an issue to be tackled on a planetary level, and not only locally as often occurred in the past. Indeed, as highlighted in this contribution, the trade in food commodities between countries and the different continents leads to a corresponding virtual water movement throughout the globe, with very significant and, sometimes, predominant flows concerning the water volumes consumed in agri-food production. This is the case for Italy where, over the last 25 years, there has been a progressive reduction in areas used for agriculture, a reduction only partially compensated for by an increase in crop yields. There followed a slight decline in the virtual water volumes used for agricultural purposes (from 72 km^3 in 1986 to 66 km^3 in 2010), accompanied by a growth in national virtual water consumption (from 111 to 121 km^3 in the 25 years considered), to a large extent, due to population growth. Summing-up, the gap between internal virtual water supply/demand widened from 39 to 55 km^3 and the differences were compensated for by a very important growth in agri-food good imports. More generally, the volumes of virtual water traded from Italy toward the other countries more than doubled from 1986 to 2010. Using virtual water as an indicator, a rather well-defined picture can be drawn. Italy appears to be a country which is increasingly abandoning its agricultural-productive role and is instead increasing its commercial one. In the near future, water resource management in Italy, just as for agricultural and land protection policies, will have to duly take into account, as much for water as for other resources, the understanding of the phenomenon and its correct interpretation, increasingly moving towards recognising the global scale of these issues. In other words, the globalisation of water has become an unavoidable fact.

Bibliography

Allan J. A. (1993). Fortunately there are substitutes for water otherwise our hydro-political futures would be impossible. In *Priorities for water resources allocation and management* (pp. 13–26). London: ODA.

Allan, J. A. (1998). Virtual water: A strategic resource global solutions to regional deficits. *Ground Water, 36*(4), 545–546.

Carr, J., D'Odorico, P., Laio, F., & Ridolfi, L. (2012). Inequalities in the networks of virtual water flow. *EOS, 93*(32), 309–310.

Carr, J., D'Odorico, P., Laio, F., Ridolfi, L., & Seekell, D. (2013). Recent history and geography of the virtual water trade. *PLOS, 8*(2), e55825.

De Fraiture. C., Cai, X., Amarasinghe, U., Rosegrant, M., Molden, D. (2004). Does international cereal trade save water? The impact of virtual water trade on global water use. Comprehensive Assessment Research Report 4. Sri Lanka: IWMI.

Falkenmark, M., & Rockström, J. (2004). *Balancing water for humans and nature*. London: Earthscan.

Falkenmark, M., & Rockström, J. (2006). The new blue and green water paradigm: Breaking new ground for water resources planning and management. *Journal of Water Resources Planning and Management, 132*, 129–132.

Hanjra, M. A., & Qureshi, M. E. (2010). Global water crisis and future food security in an era of climate change. *Food Policy, 35*(5), 365–377.

Hoekstra, A., & Chapagain, A. K. (2008). *Globalization of water: Sharing the planet's freshwater resources*. Oxford (UK): Blackwell.

Hoekstra, A., & Mekonnen, M. M. (2012). The water footprint of humanity. *Proceedings of the National Academy of Sciences of the United States of America, 109*(9), 3232–3237.

Oki, T., & Kanae, S. (2004). Virtual water trade and world water resources. *Water Science and Technology, 49*(7), 203–209.

Rosegrant, M. W., Cai, X., & Cline, S. A. (2002). *World water and food to 2025: Dealing with scarcity*. Washington, DC.: IFPRI.

Tamea, S., Allamano, P., Carr, J. A., Claps, P., Laio, F., & Ridolfi, L. (2012). Local and global perspectives on the virtual water trade. *Hydrology and Earth System Sciences Discussions, 9* (11), 12959–12987.

Verma, S., Kampman, D. A., Van der Zaag, P., & Hoekstra, A. Y. (2009). Going against the flow: A critical analysis of inter-state virtual water trade in the context of India's National River Linking Programme. *Physics and Chemistry of the Earth, 34*, 261–269.

Vörösmarty, C. J., et al. (2010). Global threats to human water security and river biodiversity. *Nature, 467*, 555–561.

World Economic Forum. (2011). *Water security: The water-energy-food-climate nexus*. Washington, DC: Island Press.

Virtual Water Trade in the Mediterranean: Today and Tomorrow

Roberto Roson and Martina Sartori

Abstract Virtual water trade refers to the implicit content of water in the production of goods and services. When trade is undertaken, there is an implicit exchange of water. In countries where water is relatively scarce, water-intensive goods are expensive to produce, so that imports normally exceed exports and the economy virtually imports water. This paper provides some estimates of virtual water trade patterns in the Mediterranean, which is an area where water is scarce, unevenly distributed, and progressively insufficient because of climate change and reduced precipitation. We analyse two cases: the current virtual water trade structure, related to trade in agricultural goods, and a future scenario, simulated by means of a computable general equilibrium model, where reduced agricultural productivity, induced by lower water availability, is taken into account.

Keywords Computable general equilibrium models · Water · Virtual water · Water scarcity · Climate change

Jel Codes C68 · D58 · F18 · Q17 · Q24 · Q54 · Q56

R. Roson (✉)
Dipartimento Di Scienze Economiche, Cà Foscari University, Cannaregio 873, 30121 Venice, Italy
e-mail: roson@unive.it

R. Roson · M. Sartori
IEFE, Bocconi University, Milan, Italy

M. Sartori
School of International Studies, University of Trento, Trento, Italy

© Springer International Publishing Switzerland 2015

M. Antonelli and F. Greco (eds.), *The Water We Eat*,
Springer Water, DOI 10.1007/978-3-319-16393-2_13

1 Introduction

Water availability is a key factor in many societies, shaping cultures, economies, history, and national identity. This is especially true in the Mediterranean, where water resources are limited and very unevenly distributed over space and time.

There is a growing concern about water resources in this region. On the demand side, during the second half of the twentieth century, water demand has increased twofold, reaching 280 km^3/year (UNEP 2006). Much of the demand comes from agricultural activities (45 % in the North, 82 % in South and East), but other industries also contribute significantly (most notably, tourism) and more competition for water resources can be easily foreseen in the near future.

Availability of water resources affects international trade and the relative competitiveness of countries and industries. In this context, "virtual water" is a useful concept to highlight the link between water consumption and trade (Allan 1993). The virtual water content of a good is defined as the volume of water that is actually used to produce that product. This will depend on the production conditions, including place and time of production and water-use efficiency. Producing one kilogram of grain in an arid country, for instance, can require two or three times more water than producing the same amount in a humid country (Hoekstra 2003).

When a good is exported, its virtual water content is implicitly exported as well. Vice versa, when one good is imported, the water used in its origin country of production is virtually imported. An origin/destination trade matrix can therefore be translated in terms of virtual water equivalent flows, allowing one to see whether a country is a net importer or exporter of virtual water, and which are its trade partners.

A large and flourishing literature on virtual water, as well as on the related concept of water "footprint", is now available (for a critical review, see Yang and Zehnder 2007). Recently, the National Geographic (2010) magazine provided a map of virtual water trade flows in the world. The idea behind the virtual water concept is not restricted to water, but applies equally well to any resource, for example carbon, so we can also discuss about "virtual carbon" trade (Atkinson et al. 2010), that is, carbon emissions generated by foreign consumption (more often named "carbon leakage").

In this paper, we estimate and analyse the current virtual water trade flows for some countries in the Mediterranean. Following Antonelli et al. (2012), we consider both the direct and the indirect (that is, associated with intermediate factors) water consumption, and we make a distinction between "green" and "blue" water.

We compare the current picture of virtual water trade in the Mediterranean with a counterfactual one, where we simulate the effects of reduced water availability. To this end, we first construct a scenario, accounting for changes in water demand and supply at the year 2050. We then estimate the associated changes in productivity for agricultural industries in some Mediterranean countries. Subsequently, we simulate the effects of changing productivity on international trade by means of a Computable General Equilibrium model of the world economy (Hertel and Tsigas 1997).

The counterfactual trade patterns estimated by the CGE model are then translated in terms of virtual water trade flows, so that an assessment can be made about how reduced water availability would affect the structure of virtual water trade in the Mediterranean.

The paper is organized as follows. In the next section, some estimates of current virtual water trade flows in the Mediterranean are presented and discussed. Section 3 illustrates how future changes in water availability, and the related impact on agricultural productivity, have been obtained. Section 4 presents the results of the CGE simulation exercise in terms of virtual water trade. A final section provides some concluding remarks.

2 Current Virtual Water Trade in the Mediterranean

To estimate current virtual water trade flows, we consider 11 countries and 3 regional economies, obtained through aggregation from the GTAP 8 database.[1] These are Albania, Croatia, Cyprus, Egypt, France, Greece, Italy, Morocco, Spain, Tunisia, Turkey, Rest of Europe (Xeur), Rest of Middle East and North Africa (XMENA), and Rest of the World (RoW). Chapagain and Hoekstra (2004) provide estimates of total water consumption for 164 crops in 208 countries. We aggregate data to the 14 regions and 7 agricultural industries of the GTAP database, and then, we make a comparison between water consumption, by crop and region, and value of production (2004). This creates an estimate of direct water usage by unit of output (in monetary terms).

The direct water usage should not be confused with the unit virtual water content, as the latter includes the water indirectly consumed through the utilization of intermediate production factors. Antonelli et al. (2012) show how virtual water coefficients can be estimated, while taking into account the input–output linkages among sectors in the economy. We apply this methodology to get the systemic virtual water consumption, per unit of output, for each agricultural product in all regions. These parameters allow translating trade flows into equivalent virtual water units.

The whole matrix of bilateral virtual water trade flows corresponding to agricultural trade flows in the 2004 GTAP 8 database is displayed in the Appendix. In the following, we illustrate some summary indicators of current virtual water trade in the Mediterranean, to highlight some of its key characteristics.

Figure 1 shows the per capita virtual water trade balance, that is the difference between total virtual water exports and imports. It can be interpreted as a measure of trade-related water dependence. Countries where water resources are scarce and expensive are generally expected to have a comparative disadvantage in water-intensive industries, resulting in net virtual water imports. However, this outcome

[1] See: http://www.gtap.org.

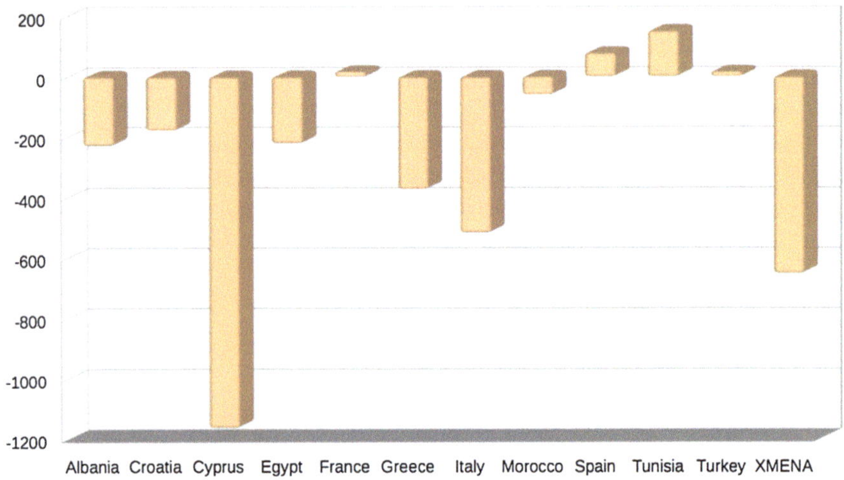

Fig. 1 Per capita net virtual water exports/balance (Mm^3)

may not necessarily emerge, because of a number of other effects at work, like market failures (severe in the water sector, where resources may be underpriced and over-exploited) and other factors affecting the overall competitiveness of an industry in the world economy.

We can see that Cyprus is the country which indirectly obtains the largest amount of water through trade in agricultural goods, followed by Rest of Middle East and North Africa, Italy, and Greece.

More information about the sources of virtual water trade can be obtained through a decomposition of flows by region of origin and destination, which is provided in Fig. 2. In this figure, each country and region is associated with a negative and a positive bar. Negative bars express imports and positive bars express exports. Each bar is split in terms of countries of origin and destination, all with a distinctive colour. The trade balance displayed in Fig. 1 is just the algebraic difference between total exports and total imports in Fig. 2.

Figure 2 highlights some interesting facts. First, some countries (e.g. Cyprus and XMENA) have a polarized trade: they import a lot but make almost no exports. Other countries, like Spain, are more balanced: they have quite large input and output flows, suggesting that they may play a role as hub in agricultural markets, possibly by importing raw agricultural goods and exporting refined, processed goods. Second, some countries have principal and different trading partners for imports and exports. Spain, but also Morocco and Italy, gets the bulk of their virtual water imports from outside Europe and the Mediterranean, exporting mainly towards central and northern Europe.

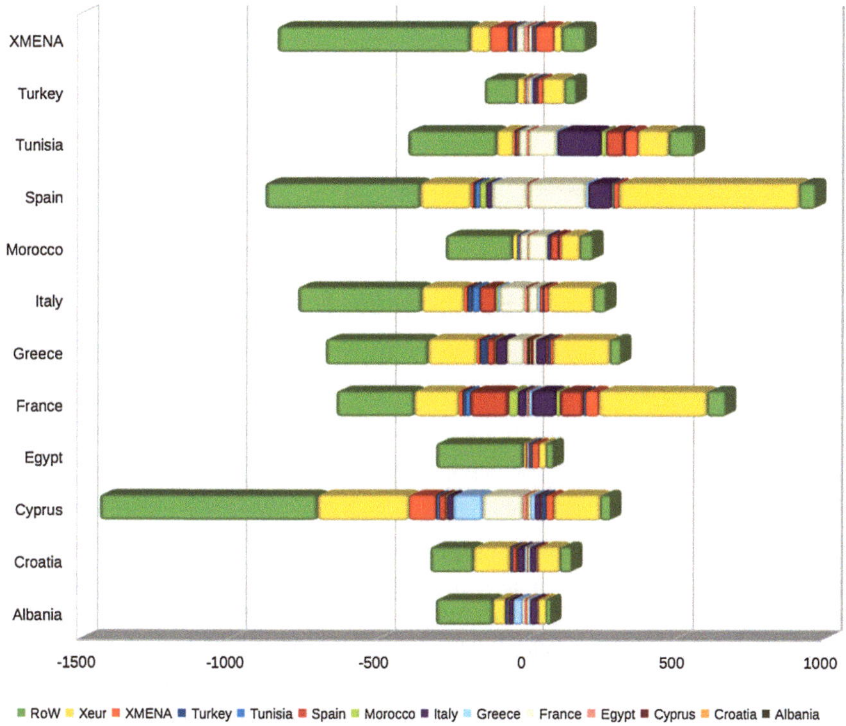

Fig. 2 Per capita virtual water trade flows in Mediterranean countries (Mm^3)

Figure 3 provides a picture of major virtual water flows on a geographical map. It displays only the largest flows, where thickness of the arrow line depends on the flow magnitude.[2]

The most significant exchanges of virtual water are found between the largest North-Mediterranean economies. France and Spain are the greatest traders of agricultural goods. An important role is also played by Italy, which is a substantial importer of agricultural products, and by some North African countries, such as Morocco and Tunisia, as well as by Turkey.

Further insights can be obtained by making a distinction between *green* and *blue* water trade flows. Green water is water stored into the soil moisture. Blue water is surface and underground water. The distinction is important because, whereas green water can be used only in agriculture, blue water can be allocated to alternative uses. Therefore, it has a greater economic potential and value. To the extent that virtual water trade is an indirect mean of saving on water resources, savings obtained on precious blue water is what matters the most.

[2]Classes are as follows: 500–1000 Mm^3, 1000–3000 Mm^3, 3000–7000 Mm^3, >7000 Mm^3.

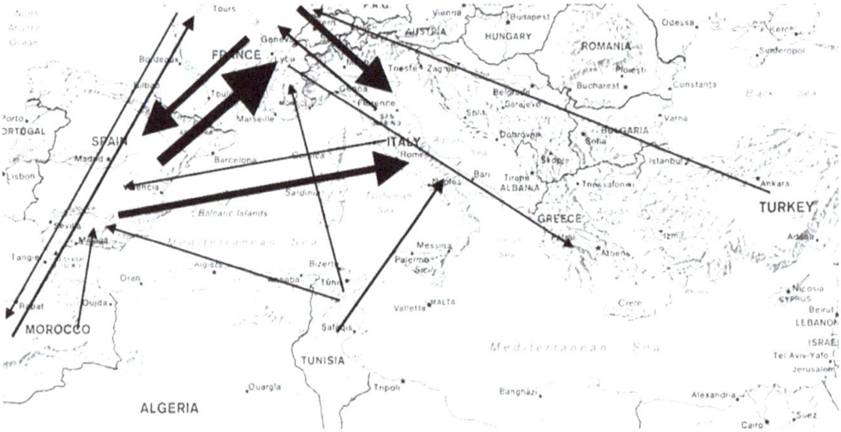

Fig. 3 Largest flows of virtual water trade in the Mediterranean

From our estimates, it is possible to assess the blue virtual water "savings" achieved by means of trade in agricultural goods. This is done by computing how much blue water would be needed if imports were instead produced domestically, while subtracting from the latter the blue water consumed to produce the exported goods. Therefore, this new balance expresses how much extra blue water would have been necessary if a country would not have been involved in international trade, at given (unchanged) levels of domestic consumption. Results are shown in Fig. 4.

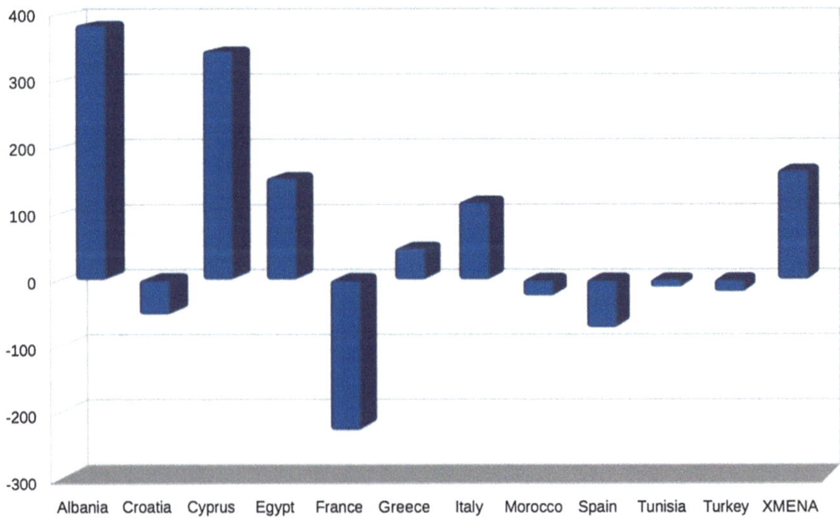

Fig. 4 Per capita *blue* water "savings" through trade in agricultural goods (Mm^3)

Much of the countries "save" on blue water by trading agricultural goods. Albania and Cyprus display the most significant savings, followed by Rest of Middle East/North Africa, Egypt, Italy, and Greece. France is the only country exhibiting an exploitation of blue water resources, but this comes quite naturally, as blue water is relatively abundant in that country.

3 A Scenario of Future Water Availability for the Mediterranean

Virtual water patterns will change in the future because international trade will. Any factor affecting the world economy, including growth, economic policies, demography and external shocks, has the potential to generate repercussions on virtual water flows.

In this paper, we focus on the prospective effects of varying water availability in the Mediterranean. Drawing upon findings of the European research project WASSERMed,[3] we consider a scenario at the year 2050, where supply and demand of water in the different countries are evaluated. Variations in agricultural productivity induced by the changing water availability are subsequently inserted into a global CGE model, to simulate the structural adjustment process for the world economy, and its implications in terms of virtual water trade. This section illustrates how the water availability scenario has been constructed.

The starting point is the current water balance for the 14 regional economies in our data set, that is the relationship between water supply (sources) and water demand (uses) at one period in time (e.g., a year). Among the supply sources, we distinguish between blue water and green water, supposing that only a fraction of total blue water is technically and economically accessible/exploitable. We estimate total water availability as the sum of accessible blue and green water, elaborating on data provided by Gerten et al. (2011).

Three uses of water are appraised: agricultural, municipal, and industrial. Green water can only be used in agriculture, where it is supplemented by blue water through irrigation or other means. Any difference between total blue water availability and consumption in the three categories above (where agricultural consumption is considered only for the part exceeding the green water stock) is interpreted either as unused water or water deliberately left for the preservation of aquatic ecosystems, which we refer to as environmental flow requirement (EFR). Water consumption by agriculture in the baseline (2000–2005) has been estimated using data from Chapagain and Hoekstra (2004). Municipal and industrial consumption has been obtained from the FAO—AQUASTAT database. Figure 5

[3]Water Availability and Security in Southern Europe and the Mediterranean Region (WASSERMed) is a research project funded by the European Commission in the 7th Framework Programme (contract no. 244255). For more information, see http://www.wassermed.eu.

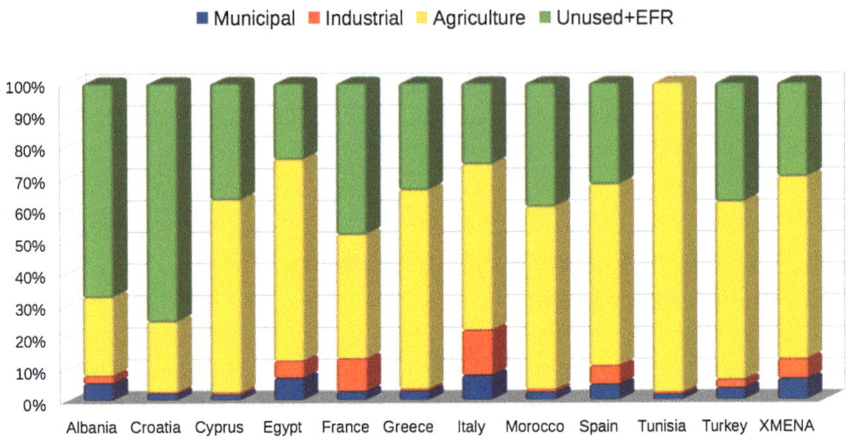

Fig. 5 Composition of (blue) water consumption in the baseline (2000–2005)

shows the composition of water demand among different usage classes for the Mediterranean countries considered in this study.

To assess how much water will be available for agriculture in 2050, values for all supply and demand components have been projected to the future. On the supply side, total water availability has been estimated starting from data about future climate conditions, produced in WASSERMed from a set of regional and global circulation models. The climate scenario suggests that precipitation will generally decrease in the Mediterranean in the period 2000–2050, particularly in France (−13 %), Morocco (−18 %), and Tunisia (−10 %). The average temperature is expected to increase of about 2 °C. Climate scenario affects the future total water availability by means of changes in precipitation and temperature. Elaborating on these data,[4] we predict a significant drop in water supply for France (−34 %), Italy (−14 %), Turkey (−11 %), Croatia (−10 %), Spain (−8 %), whereas countries in southern Mediterranean would be much less affected, because their water resources are relatively more independent from local climate conditions.

On the demand side, the various components have been projected using different assumptions and methodologies. For water consumption in agriculture, our reference point is a hypothetical situation in which agricultural production volumes stay unchanged. Of course, this is not meant to be a realistic scenario, but only a reference benchmark. Even with constant production levels, however, water demand would increase, because of higher temperature and evapotranspiration, of about 10 %.

---

[4]We considered that changes in precipitation do not automatically translate into changes in total water availability. Details and data of the elaboration process are obtainable from the authors upon request.

Municipal consumption, that is water for human drinking, washing, etc., is generally assumed to follow demographic changes. Population projections have been taken from the World Population Prospect (United Nations Secretariat 2010), which devises a very strong growth of population in the Middle East. In addition, for some developing countries in our set, we consider the possibility that municipal water demand could increase more than proportionally, to reflect improved access to sanitation and freshwater.

Industrial consumption is assumed to increase at a rate equal to 1/3 of the national income growth. GDP forecasts have been derived from World Bank Statistics.[5] The lower rate for industrial water consumption is intended to account for the changing composition of the national income, with a lower share for manufacturing industries, as well as improvements in efficiency.

In addition to water consumption, water resources may be needed to preserve a number of natural environments. The environmental flow requirement expresses this "pseudo-demand" for water as a share of total runoff, so we logically extend the notion to our estimates of blue water availability. The EFR concept itself is a rather elusive one, as there is no fixed threshold value for environmental preservation and much depends on collective evaluation. We look at the literature (Korsgaard 2006; Hirji and Davis 2009) to select some "reasonable values" for the EFR, which in this context means the share of blue water resources that should be set aside for effective protection of the environment.

We regard the EFR not as a constraint but as a policy variable. In other words, national governments may or may not be willing to save water for environmental purposes. In our numerical experiments, we assume that all countries in the European Union[6] must comply with strict environmental regulation, so that the EFR share of (blue) water cannot be made available for consumption. Non-EU countries, on the other hand, are assumed to have more degrees of freedom, so they may opt not to comply with EFR requirements. In our scenario, we assume that non-EU countries do not comply with EFR requirements.

When municipal and industrial consumption and possibly EFR are subtracted from total blue water, what is left is water potentially available for the agricultural sector, supplementing green water. This "water potential" can be compared with estimates of agricultural water demand at fixed production levels. If potential water exceeds water demand, agriculture is not water constrained, at least if current production volumes do not significantly increase. Otherwise, (blue) water delivered to agriculture must be cut by a certain amount. Table 1 presents our estimates of reductions in water available for agriculture in 2050.

Five economies are found to have insufficient water resources, at the year 2050, to sustain current production levels in agriculture. Because of lower precipitation and higher temperature, France and Italy will be affected by a drop in water resources, with a larger impact for agriculture in Italy, because relatively more blue water is

[5]http://www.data.bank.org.

[6]The group includes Croatia and Turkey as accession countries.

Table 1 Reductions in water available for agriculture in 2050 (only affected regions)

Egypt	−12 %
France	−15 %
Italy	−19 %
Tunisia	−11 %
XMENA	−8 %

used there. Tunisia, Egypt, and Rest of Middle East/North Africa will also lack water for agriculture, but for different reasons. In Egypt and the Middle East, the climate change impact on total water availability will be negligible, as in this area much of the water is imported, pumped from the ground, desalinated, or recycled. However, non-agricultural water uses are expected to grow at a significant rate. Tunisia is a special case, since water resources are overexploited already in the baseline (see Fig. 5). Any further increase in water demand would not be sustainable.

We analyse how reductions in water availability could affect agricultural productivity. To this end, it is important to take into account that (i) each country has its own mix of agricultural products, and (ii) crops may differ in terms of sensitivity to water shortages. As in the GTAP database, we consider seven classes of agricultural products: wheat, cereals, rice, vegetables and fruits, oilseeds, sugar, and other products. For each crop group in each country, a "water elasticity" parameter has been estimated, accounting for both the physical characteristics of the crop and the overall efficiency of the water delivering system. Using these parameters, changes in water availability for agriculture have been translated into changes in agricultural productivity by sector, as reported in Table 2.

On average, the impact on agricultural productivity is very high for Egypt, high for France and Italy, and medium for Tunisia and the Rest of Middle East/North Africa.

4 Future Virtual Water Trade in the Mediterranean

Estimates of Table 2 have been used to shock exogenous productivity parameters in a computable general equilibrium model of the world economy (Hertel and Tsigas 1997). A CGE model is a very large nonlinear system, which provides a systemic

Table 2 Reduction in agricultural productivity by sectors and by region

	Wheat (%)	Cereals (%)	Rice (%)	Veg. and fruits (%)	Oilseeds (%)	Sugar (%)	Other crops (%)
Egypt	−17.03	−17.03	−21.72	−21.72	−22.00	−22.00	−20.25
France	−23.54	−23.54	−16.28	−16.28	−10.61	−10.61	−16.81
Italy	−16.96	−16.96	−14.63	−14.63	−10.00	−10.00	−13.86
Tunisia	−1.62	−1.62	−7.67	−7.67	−4.01	−4.01	−4.43
XMENA	−1.12	−1.12	−5.27	−5.27	−2.76	−2.76	−3.05

and disaggregated representation of national, regional, and multi-regional economies. The system includes market clearing conditions and accounting identities, to account for the circular flow of income and inter-sectoral linkages inside the whole economic system.

A simulation exercise entails comparing two equilibria for the global economy, in which all markets clear, before and after the variation of some exogenous parameters (in our case, multi-factor productivity in a set of agricultural industries). The model output includes all the main macroeconomic variables, such as nominal and real GDP, consumption and production levels, relative prices for products and primary factors. To summarize the overall macroeconomic impact of the simulated variation in agricultural productivity, Table 3 reports the estimated percentage variations for the national real income (which is a measure of aggregate welfare) for all regions in our set.

Not surprisingly, lower productivity in agriculture, induced by reduced water availability, generates negative consequences in terms of national income for most Mediterranean countries. The magnitude of the loss depends on the amount of the productivity shock, but also on the share of agricultural activities in the economy. Egypt is the country which is hurt the most, as the model estimates a fall of 7.24 % for real income. Significant reductions in GDP and welfare are also estimated for France, Italy, and Tunisia, which are the other water-constrained countries in our exercise. Three countries get (slight) benefits: Morocco, Spain, and Turkey. This is not because of improvements in productivity (which is unchanged there) but because of enhanced relative competitiveness vis-à-vis trading partners and competitors (a second-order general equilibrium effect).

Table 3 Estimated variations in real national income

	Var. (%)
Albania	−0.17
Croatia	−0.05
Cyprus	−0.02
Egypt	−7.24
France	−1.83
Greece	−0.01
Italy	−1.77
Morocco	0.10
Spain	0.04
Tunisia	−1.42
Turkey	0.01
Rest of Europe	−0.03
Rest of MENA	−0.32
RoW	−0.003

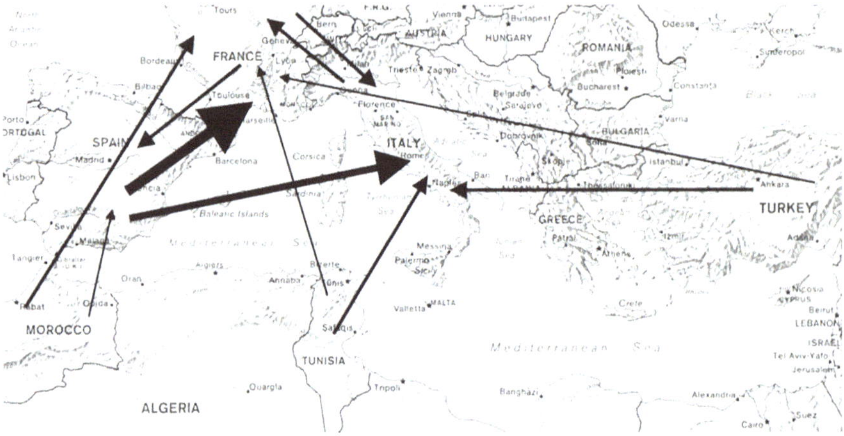

Fig. 6 Largest flows of virtual water trade in the Mediterranean in the counterfactual scenario (Mm^3)

The output of the CGE computer simulation comprises counterfactual estimates of trade flows. Using the same procedure applied for actual trade flows, illustrated in Sect. 2, it is possible to estimate virtual water flows for the scenario under consideration. Figure 6 is analogous of Fig. 2 and displays the most significant flows of virtual water between Mediterranean countries.

The most notable difference between Figs. 2 and 6 is that the amount of virtual water flowing from France toward Italy, Spain and, to a lesser extent, Morocco decreases significantly. On the other hand, imports of virtual water by France increase somewhat. The reason is easily found by looking at Fig. 7 (corresponding to Fig. 1 in Sect. 2), presenting estimates of the virtual water trade balance.

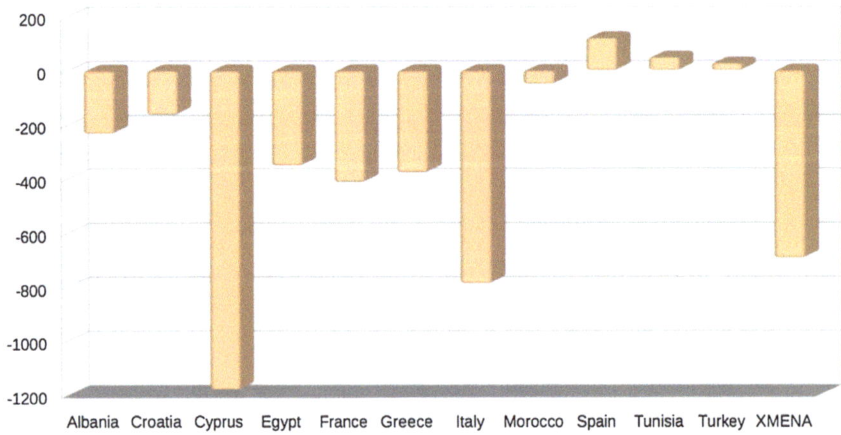

Fig. 7 Per capita net virtual water exports (balance) in the counterfactual scenario (Mm^3)

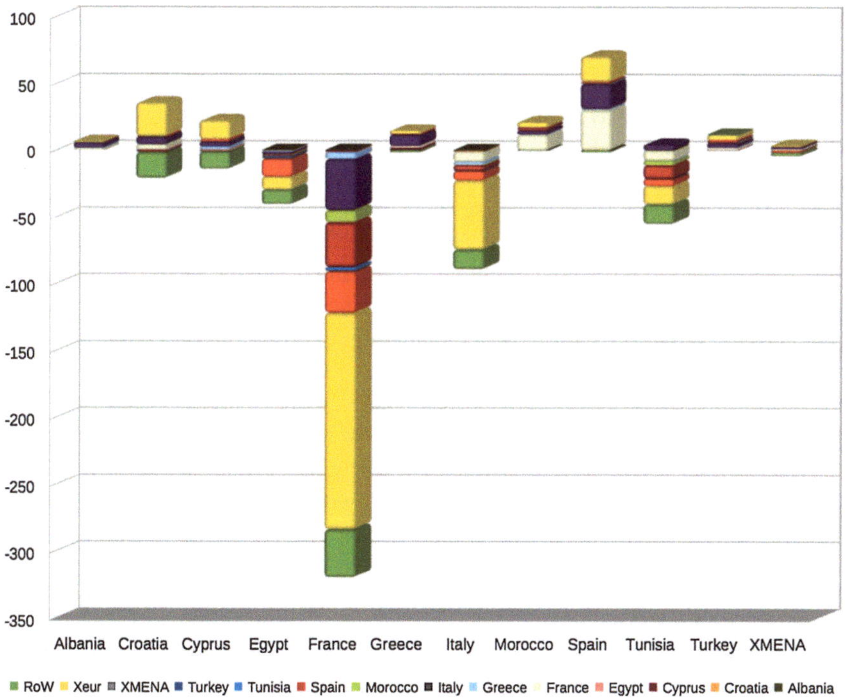

Fig. 8 Variation in per capita virtual water flows (Mm^3) for each Mediterranean economy

In the current baseline (Fig. 1), the French virtual water trade balance was slightly positive. In the counterfactual scenario (Fig. 7), the balance gets negative, because the agricultural sector in France becomes much less competitive. Italy and Tunisia also deteriorate their virtual water trade balance, with less water flowing towards Spain and Italy, respectively. On the other hand, Morocco and Turkey grow in terms of virtual water exports.

Figure 8 shows *from where* the additional water imports come from, or *to where* additional water exports are directed. As such, this figure does not completely correspond to Fig. 2. We can see that France and Italy get most of the extra water imports from central and northern Europe. On the other hand, Spain increases its virtual trade exports towards the latter two countries.

5 Concluding Remarks

This paper has provided some estimates of virtual water trade patterns in the Mediterranean. Virtual water trade follows conventional trade in goods (in this case agricultural products), whereas no water is physically exchanged. As international trade is a powerful mechanism for improving the allocation of economic resources, including water, the virtual water paradigm is just one way of looking at the potential benefits of international trade from a "water perspective".

The Mediterranean is an area where water is scarce and unevenly distributed. Potential water demand exceeds supply in many Mediterranean countries, and problems are likely to be exacerbated in the future, because of climate change and reduced precipitation.

In this work, two cases have been considered: the current virtual water trade structure, related to trade in agricultural goods, and a future scenario, simulated by means of a computable general equilibrium model, where reduced agricultural productivity, induced by lower water availability, is taken into account.

Analysis of current virtual water flows reveals that most countries are net importers of virtual water, thereby realizing sizeable "savings", particularly of precious blue water resources. Much of the intra-Mediterranean virtual water trade occurs between the largest northern economies (Spain, France, Italy) but, in per capita terms, the country which gets the largest amount of virtual water from abroad is Cyprus.

This picture will likely change in the time ahead, because the evolution of the world economy, as well as of international trade, will ultimately be reflected in varying virtual water flows. A simulation exercise has been performed in this paper, where we abstract from the many possible factors affecting future trade patterns, focusing instead only on the possible consequences of reduced water availability in the Mediterranean.

We found that both northern and southern countries will be affected by water shortages, although for different reasons. In the north, increased temperature and reduced precipitation will lessen water stocks. In the south, the driving factors will be demographic and economic development. Implications of this scenario in terms of virtual water entail reduction in intra-Mediterranean trade and increases in virtual imports from central and northern Europe, as well as from the rest of the world.

Acknowledgments This study has been partly funded and realized in the context of the EU FP7 project WASSERMed (Grant Agreement Number 244255, http://www.wassermed.eu). Ana Iglesias and Sonia Quiroga helped us in several ways during the development of this work, most notably in the definition of the water availability scenario. Marta Antonelli provided essential data on non-agricultural water consumption. The usual disclaimer applies.

Appendix

See Table 4.

Table 4 Baseline virtual water trade flows (Mm^3)

	Albania	Croatia	Cyprus	Egypt	France	Greece	Italy	Morocco	Spain	Tunisia	Turkey	XMENA	XEur	RoW	Tot. exp.
Albania	0.0	1.0	0.1	0.4	12.9	13.4	77.4	0.1	3.6	0.1	6.9	2.2	78.5	51.5	248.2
Croatia	7.4	0.0	2.8	3.2	18.5	4.4	103.9	0.7	7.5	0.2	3.9	9.3	341.1	188.1	691.0
Cyprus	0.1	1.1	0.0	1.1	5.9	11.0	15.7	0.0	2.8	0.0	11.0	19.9	120.8	27.6	217.1
Egypt	22.0	4.9	9.1	0.0	120.3	114.7	399.9	40.8	119.6	50.9	164.4	1545.9	1584.4	1711.8	5888.9
France	22.4	17.0	105.4	78.5	0.0	592.8	4847.4	719.9	4656.4	215.1	111.9	2600.1	21,473.3	3800.5	39,240.8
Greece	95.9	18.0	78.1	19.3	82.3	0.0	373.3	2.2	58.3	5.1	73.9	81.1	1997.9	398.8	3284.1
Italy	47.6	118.4	16.3	17.0	1684.3	409.3	0.0	13.2	721.5	20.5	86.6	764.9	8595.7	2402.4	14,897.7
Morocco	1.0	5.8	1.6	13.9	1893.0	22.8	317.0	0.0	847.8	22.1	10.9	150.0	1846.1	1342.1	6474.1
Spain	3.6	65.5	19.5	15.1	7599.3	260.9	3158.4	104.6	0.0	58.8	78.3	633.9	24,254.1	2199.5	38,451.5
Tunisia	1.3	5.5	4.2	8.2	871.6	26.3	1380.1	149.4	569.1	0.0	27.0	403.3	975.8	837.3	5259.1
Turkey	28.2	26.3	2.3	79.8	555.9	242.0	957.8	22.2	260.0	36.9	0.0	1020.1	4835.3	2641.4	10,708.2
XMENA	2.6	9.5	74.0	226.8	836.0	112.8	693.4	37.8	368.1	60.6	294.3	6694.8	2707.7	8846.3	20,964.7
XEur	124.1	564.4	241.2	340.9	8945.6	1737.4	8234.2	411.3	6856.3	513.3	1516.3	7077.4	62,861.7	19,109.8	118,533.7
RoW	607.0	681.9	574.4	20,061.9	15,608.5	3687.7	24,074.3	6755.5	20,876.7	2853.7	7301.3	71,166.7	143,929.1	763,887.2	1,082,066.1
Tot. imp.	963.2	1519.5	1129.1	20,866.1	38,234.1	7235.6	44,632.6	8257.8	35,347.9	3837.2	9686.5	92,169.7	275,601.7	807,444.4	

References

Allan, J. A. (1993). Fortunately there are substitutes for water otherwise our hydro-political futures would be impossible. In *ODA, priorities for water resources allocation and management* (pp. 13–26). London: ODA.

Antonelli, M., Roson, R., & Sartori, M. (2012). Systemic input-output computation of green and blue virtual water 'flows' with an illustration for the mediterranean region. *Water Resources Management*. doi:10.1007/s11269-012-0135-9

Atkinson, G., Hamilton, K., Ruta, G., & van der Mensbrugghe, D. (2010). *Trade in 'virtual carbon'—empirical results and implications for policy*. Policy Research Working Paper 5194. Washington, D.C.: The World Bank.

Chapagain, A. K., & Hoekstra, A. Y. (2004). *Water footprints of nations*, Vol. 2: Appendices. Value of water research report series No. 16, UNESCO-IHE Delft.

Gerten, D., Heinke, J., Hoff, H., Biemans, H., Fader, M., & Waha, K. (2011). Global water availability and requirements for future food production. *Journal of Hydrometeorology, 12*, 885–899.

Hertel, T. W., & Tsigas, M. (1997). Structure of GTAP. In T. W. Hertel (Ed.), *Global trade analysis: Modeling and applications*. New York: Cambridge University Press.

Hirji, R., & Davis, R. (2009). *Environmental flows in water resources policies, plans, and projects: Case studies*. Washington, D.C.: The World Bank Environment Department.

Hoekstra, A. Y. (2003). Virtual Water: an Introduction. In A. Y. Hoekstra (Ed.), Virtual water trade. In *Proceedings of the International Expert Meeting on Virtual Water Trade*. Value of Water Research Report Series No. 12. The Netherlands: IHE Delft.

Korsgaard, L. (2006). *Environmental flows in integrated water resources management: Linking flows, services and values*. Institute of Environment & Resources: Technical University of Denmark.

National Geographic. (2010). *Water—our thirsty world,* Special Issue (US Ed.), April 2010. Washington, D.C.: National Geographic Society.

UNEP—Plan Bleu. (2006). *Facing water stress and shortage in the Mediterranean*, Blue Plan Notes no. 4, October 2006.

United Nations Secretariat. (2010). *World population prospects: The 2010 revision.* Department of Economic and Social Affairs. http://esa.un.org/unpd/wpp/index.htm

Yang, H., & Zehnder, A. (2007). Virtual water: An unfolding concept in integrated water resources management. *Water Resources Research, 43*(12), 1–10.

An Economic Approach to Water Scarcity

Antonio Massarutto

1 Water as an Economic Good

A limitation—but also an advantage, depending on the viewpoint—of economic evaluation is to refrain from any a priori criteria, value judgments and ethical parameters. For an economist, the concept of "value" always refers to an objective dimension, which descends from the fact that a given action determines a variation in the utility of some individuals. Evaluation actually concerns the measurement of how individual or collective utility change, as a consequence of the action that is analysed. Therefore, for an economist, ideas such as the "ecological footprint", "water footprint", "virtual water" and similar carry a neutral meaning, since a high consumption of water (or of anything else) has not, in itself, a negative or positive connotation. What we need to know instead is the extent to which a given use of a resource affects other possible alternative uses of the same resource that are on their own "valuable", that is, generate positive utility: Is anybody's utility negatively affected, and how? Does use *x* prevent or limit the use *y*?

Clearly, it is intuitive that using a lot of water where it is available in sufficient quantities to satisfy all uses, including anthropogenic and ecosystemic ones, cannot be considered a negative value.

The case of water is also complicated by the fact that it involves, on the one hand, a good which is cyclically renewed and, on the other hand, is "flowing", with limited possibilities for storage and conservation. In other words, it must be considered that the use does not necessarily negate the possibility of an alternative use (the same water, returned to the cycle, will be once again available).

At the same time, non-use does not necessarily mean a saving, since water that flows without being used will sooner or later end up in the sea.

A. Massarutto (✉)
IEFE, Bocconi University and University of Udine, Udine, Italy
e-mail: antonio.massarutto@unibocconi.it; antonio.massarutto@uniud.it; antonio.massarutto@dse.uniud.it

© Springer International Publishing Switzerland 2015

M. Antonelli and F. Greco (eds.), *The Water We Eat*,
Springer Water, DOI 10.1007/978-3-319-16393-2_14

Therefore, the important elements for an economic analysis concern not so much the consumption, but the concentration of the consumption over time and the possible occurrence of conflicts among competing users. The resource availability has to be evaluated instant by instant. An important impact, in economic terms, can take place, for example, because in that exact moment use "*x*" prevents or makes use "*y*" more difficult, or because it draws on a stock (e.g. groundwater, a lake) which denies someone else the option of using that same stock in the future.

Claiming that water is an "economic good" has nothing to do with its transformation into a commodity, namely a marketable good traded for a price. However, it also concerns the need to calculate in the costs all the sacrifices that a given use involves. Sacrifices that can be broadly divided into three categories:

- financial costs: these are the economic resources necessary to make the water available (e.g. withdrawal, transportation, treatment) and for its recovery after use, returning it to the environment (sewerage/drainage, purification, sludge treatment, etc.);
- scarcity costs (sometimes defined as "resource costs"): these are the alternative economic values which are sacrificed when a certain use hinders or prevents another. For example, if the water use by farmer *x* prevents its use by farmer *y*, the scarcity cost will be equal to *y*'s economic production value (equal to the market value of his output, after production costs);
- environmental costs (sometimes defined as "negative externalities"): these are the value of the ecosystem, paysage or similar elements where a given use of the resource has an impact. For example, the pollution caused by the release of contaminated water could result in the reduction of a water body's ecological quality, depriving it of certain environmental functions.

As can be seen, we are dealing with *contingent* cost categories, in that they cannot always be verified and cannot be easily associated with a given use or impact. A resource cost arises not only because there is someone who uses water, but also because there is someone else who would like to use it in their place. An environmental cost is quite difficult to correlate to the quantity of water withdrawn, as it depends, instead, on when, where and how it is withdrawn (and after released).

The literature regarding water footprinting and "virtual water" has certainly raised the awareness that not all uses are the same and that before establishing if a certain use results in a footprint or not it is necessary to distinguish how it occurs. For example, by distinguishing between the consumption of "blue" water (surface water or groundwater) and "green" water (rainwater contained in the soil as moisture), or, in the case of the former, depending on the renewability of the resource used, or by distinguishing "dissipative" or "non-dissipative" uses—where dissipation is, basically, reducing or eliminating availability for a given period of time.

However, to be more in line with the economic impact definition, these indicators are still much too general, in that withdrawing the same quantity of water from the same source using the same method does not necessarily impact in the same way (for further information, see in this volume "Not all drops of water are the same" by M. Antonelli and F. Greco). Therefore, these indicators are more useful for a quick overview and to identify the regions where it is more likely that a situation of stress may be identified, but they are not, however, enough in themselves to identify the situation neither to identify critical situations, nor to guide policymakers and recommend actions. Therefore, an in-depth ad hoc study for each context is necessary, both to accurately gather information on what are the stress factors (which sectors and which uses suffer from stress, and what ecosystem elements are damaged) and to identify the most suitable actions to be undertaken.

2 When Water Is Not Enough

In a certain sense, speaking of water "scarcity" is incorrect. The water available on our Earth enormously exceeds every reasonable future demand we could make on it. Of course, on a smaller territorial scale and for more limited periods of time this is less true. Yet technology literally allows us to have available any quantity of water that we may want to use. The cost of desalinating sea water is about 50 euro cents per cubic metre—a high cost, but not impossible to meet, as far as the value of the water—which descends from the market price of goods that can be produced from that cubic metre, whether they be agricultural, tourism or energy products, is higher than that cost.

And yet, the international agencies of the United Nations or the OECD have signalled the increasing risks of water stress for anthropic communities or for water ecosystems. To understand this apparent paradox, we must specify better what we mean when we maintain that water is scarce.

Scarcity cannot be measured as an absolute, but always in relation to the alternative actions that can be carried out to remedy its (temporary and local) unavailability. In other words, if here and now I do not have the water at my disposal that I wish to use, then the alternatives available to me are many.

One, of course, is to give up using it (this could be a sacrifice for me, no access means no use: but in the end I might survive anyway, if the use destination is not actually vital). Another is to move to where there is water available (this would also be a sacrifice, with all the associated costs, material and immaterial, that this move involves, whether it be temporary or permanent, and so on). Still another is to invest in technology to access the alternative resources (e.g. build an aqueduct to carry water from a place where it is available). In this case, the sacrifice will be the costs associated with the building and maintenance of the infrastructure. Yet again, I could ask someone who is using the resource denied to me to give up using it in exchange for a payment (the sacrifice would be the amount I would have to pay).

Obviously, I will choose the best alternative *for me* (in relation to the sacrifice I must make). In other words, I will always have to compare the "demand"—namely the calculation of how worthwhile it is to use the water, what are the benefits I can expect from its use—with the "supply", that is, the costs (financial, environmental, scarcity) that must be met to access it. When the cost is more than the benefit, using the water—or that water—would become economically irrational.

It is also evident that the abstract desirability to use the water must always take into account the costs. If the cost of the water is zero—that is, if it were available every time it is requested, in the quantity and quality desired—then the demand could be basically quite unlimited. In fact, there will always be an irrigable surface to be added—a new town, a new tourist village or new fountain to be built.

To define scarcity, in economic terms, the cost must be higher than zero. When the cost exceeds zero, only an economic calculation can evaluate if or what sacrifices should be made where its use is involved and allocate the available resources amongst the eventual competing uses.

Therefore, the economic idea of scarcity does not rely on the fact that the resource is more or less plentiful, but if there are mutually exclusive solutions present and levels of sacrifice which must be chosen from.

The issue becomes a little more complicated if we consider that not all those who make sacrifices are able to reason in this way. For a poor community in Sub-Saharan Africa, for example, many of the alternatives are excluded, as they have no purchasing power to allow them any technological solutions, while "relocation" may carry enormous human and social costs when this involves a mass migration. If we consider that access to water is a fundamental right, every "sacrifice" that involves giving up satisfying that right is an unacceptable cost. Furthermore, often the sacrifice is not borne by those economic players able to carry out a rational calculation of the opportunities, but falls, instead, on those subjects which, by definition, cannot bear the cost (the ecosystem, other species, future generations).

However, even these can be adequately taken into account in an economic reasoning, for example, by attributing an infinite value to the non-negotiable aspects of the sacrifice.

Figure 1 illustrates the question. In every context, water availability is limited, but could be increased with a certain cost involved (high, but in general, finite). Yet, bearing this cost would not be desirable if the value linked to use—and represented by the social demand—is not high enough. A water stress situation is usually

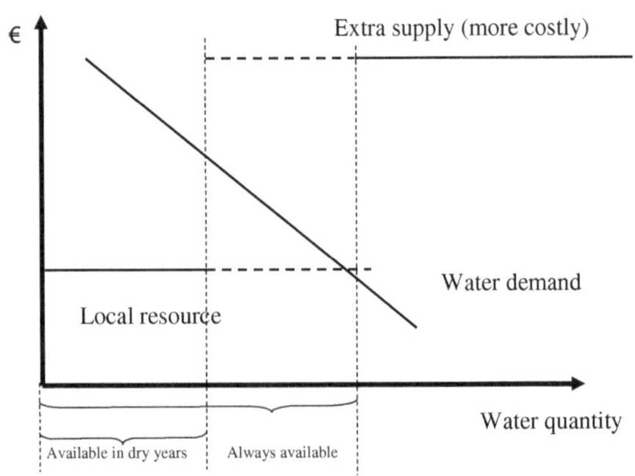

Fig. 1 The scarcity dilemma. *Source* The author

identified when the pressure on the local resources is high (the demand exceeds the quantity which the local system is able to generate at a "normal" cost, but there are not enough resources available to "make the leap" to a more costly solution).

To complicate matters further, we can see that as availability varies according to the season, stress can be verified, or not (and with more or less frequency). In the situation represented in the figure, for example, there is no stress present in normal years (the demand is met by the resources normally available), while, instead, during dry years, the supply becomes insufficient. In similar situations, some subjects, thus, remain unsatisfied. The management system could draw up rationing measures (prices, regulations), or leave it to be dealt with case by case, with the strongest prevailing, or first in first served.

3 Irrigation in Italy: A General Overview

The reasoning carried out above can also be found in an analysis of the irrigation system in Italy—a rather fitting example.

It is well known that, in Italy, the main use of water is in agriculture. Irrigation alone involves more than a half of the anthropogenic withdrawal of water. However, we should remember that these data are only an estimate and that reliable statistics are still not available, especially concerning direct water withdrawals from wells.

For an overall picture, we must turn to the dated but still valuable study conducted by the IRSA-CNR (Water Research Institute—National Research Council) in 1999, which estimated in a little less than 20 km^3 the annual withdrawal of water for agriculture compared to about 40 km^3 of total use. Besides the enormous uncertainty that surrounds these figures, the predominant weight of agricultural use is clearly evident. The irrigated surface area involves 2 million hectares, 80 % in the northern regions (*ibidem*). Most of the demand for irrigation is met by collective systems (reclamation consortia), but a substantial number of farmers use, either as an alternative or additional to collective systems, their own catchment systems (wells, lakes, etc.). On the other hand, Italy is a country blessed with water—the annual resources available are estimated at 52 km^3 (*ibidem*). If this figure is true for the whole country, then it is even more so for the northern regions. The Alpine chain, from this point of view, plays an important role as a "tank"—the snow cover, glaciers, permafrost (i.e. the ice found in the deeper soil strata), the large sub-Alpine lakes and the lowland groundwater represent a water system that not only has a plentiful annual availability, but also benefits from a natural runoff regulation during the different seasons.

Therefore, the huge quantity of water used in the north for irrigation purposes reflects the relatively high availability. A further breakdown of the figures reveals that a substantial percentage of the already high use of irrigation water in the north is for rice cultivation, where irrigation occurs at a moment in the season when competition with other users is limited (spring). The fact remains that the remaining quantities for the more traditional irrigated crops (cereals, especially maize, orchards, market gardening) would already be enough in themselves to determine a high percentage of water use.

Irrigation is of fundamental importance for all those crops where the water requirement—defined by the evapotranspiration coefficient—is not adequately offset by the natural soil moisture. Northern Italy, from this point of view, has a unique feature—its climate is markedly Mediterranean (low rainfall in the summer), but the runoffs are fed by the snowmelt and by the buffers represented by the lakes and underground water. Therefore, a relatively high water requirement can be satisfied, at quite reasonable costs, making use of this runoff. Moreover, since medieval times, the possibility of controlling the runoff using inexpensive systems (gravity water canals, spring water, etc.) has allowed for widespread agricultural development.

Irrigation has played a crucial role in agriculture in the global market networks, guaranteeing a regular production being less governed by seasonal trends. The lower risk linked to seasonality allows for the selection of more specialized crops and a higher productive yield. The *Made in Italy* food production chain, such as that for *Parmigiano Reggiano* to give just one example, relies heavily on the regular supply of animal feed, originating from irrigated crops, since production guidelines require that animals are fed with forage produced *in loco*. The recent spread of irrigation in the area of the vineyards sector is another example—the large investments made by companies in creating a wine production chain aimed at quality excellence makes the companies much more vulnerable to seasonal production fluctuations. Irrigation, with its controlled water supply, results in reducing the climatic risks and, therefore, ensuring a lower risk rate.

On the other hand, the fact that the water is plentiful and inexpensive has encouraged the development of an irrigation model that is quite lacking in terms of technology, which discourages the introduction of water-saving technique favours an efficient water use or water saving (so-called water saving technologies).

Most of the networks (more than 2/3 are in the National Irrigation Atlas published by INEA) are "free surface flow" or, however, non-pumped. This means that the water can be channelled making use of gravity, but the allocation methods for the different crops are inflexible, based on preset rotations, with no possibility for a real-time supply for the more vulnerable crops.

The paradoxical effect of this model is that enormous amounts of water are used to irrigate low value-added crops, risking, in the meantime, not having enough water available for the high value-added crops. The latter, end up "queuing", both from a geographical and seasonal point of view, to access the resource.

Thus, it should not be surprising to find in the south—generally less plentiful in water, though in relative terms suffering less damage compared to other situations clearly less favoured by nature—more modern and efficient water systems. These include water saving technology and allocation and supply methods.

4 An Economic Evaluation of Drought Costs and Possible Answers: The Po River Basin

The Po river basin, for the above-mentioned reasons, is a textbook case regarding the economic and environmental problems linked to water use in agriculture. It is a high-risk water area but, at the same time, marked by an extremely intensive use of the resources available. If the availability per capita is more than 3000 m^3/in. (the usual threshold to identify stressed regions is 1700 m^3/in.),[1] the exploitation rate is, conversely, extremely high, reaching almost 40 % of the average resources theoretically available. There is an extremely intense water use for hydroelectric energy production, also supported by an important artificial water body system. Reaching the plains, the large sub-Alpine lakes collect and control the runoff. From here, the main Lombardy and Piedmont irrigation networks spread out, intercepting the water before it percolates into the permeable subsoil of the high plains. The low plains are fed by the resurfacing of the water which then flows into the main course of the Po. The other north-eastern basins, from South Tyrol to Isonzo, display similar features. Instead, in the Apennine range, the lack of large glaciers and lakes results in a more irregular runoff. It is clear from this rapid picture that while the different water uses are in competition, they also complement each other.

80 % of the non-energy water use is concentrated in agriculture. The plentiful availability in normal years encourages an agricultural model particularly dependent on irrigation and the prevailing choice of irrigated crops of a medium–high added value, such as maize, rather than more resistant crops with a lower added value. What makes the system particularly susceptible is the inflexibility of the organizational model. In difficult seasons, this means that for many subjects, the alternative is, quite simply, to give up using water. In fact, if the crop choice is made at the beginning of the season, based on an average availability, then at the moment this availability drops below normal, or the demand is above normal (e.g. due to the effects of a very hot season or if the rainfall is below normal), many subjects find themselves unable to remedy the problem. Just as serious, obviously, is the situation of those farms which have made investments in multi-annual cycles (orchards, vineyards).

Let us take as an example what happened in 2003, a year that can be considered as paradigmatic for a stress situation. A winter with low snowfall, followed by a spring of quite low rainfall, resulted in a water level in the large lake basins very much lower than normal. This singularly dry season encouraged many agricultural companies, especially in the lower basin areas, to heavily tap into the groundwater. The result was a runoff into the main Po water course very much below the norm.

[1]This threshold is based on the Falkenmark and Lindth index, proposed in 1976 and adopted by the United Nations and the main multilateral institutions such as the OECD and the World Bank. (see, e.g. the OECD (2008)).

The large power plant of Porto Tolle, near Rovigo, which uses water from the river for its cooling system, was forced to shut down repeatedly during the month of June, resulting in a major energy shortage in the grid and the need to forecast blackout periods for many users. In order to avoid even worse consequences, the Po Basin Authority drew up an extraordinary plan that requested, amongst others, the release of the maximum quantity available from mountain hydroelectric water bodies and a constant reduction in the water use permitted (−10 %). Thus, a sufficient quantity of water was maintained in the river so as to avoid closing down the Porto Tolle plant, a cornerstone in the national energy network. On the other hand, the cost was a steep decline in agricultural production and a potential reduction in energy generated by the upstream hydroelectric stations.

Therefore, essentially, a situation such as that in 2003 exemplifies the emergence of a potential conflict between alternative water uses, where in normal conditions this does not exist as supply is sufficient for all (including the ecosystem).

In order to quantify the key economic variables, a simulation model was created (see Massarutto and de Carli 2009). Different strategies were examined involving different degrees of intervention in reducing the allowable withdrawals and different hypotheses for the reallocation of the supply amongst the users, favouring those where the water use generated a higher benefit. The model takes into account the fact that the subject that bears the main cost burden resulting from water scarcity, for example, the farmer who finds himself without water, can, in turn, transfer these effects to other subjects. For example, if because of the lack of water the agricultural production declines and the price rises, it will be the consumers who suffer the consequences. For the farmer, the effect is unclear, from the moment that the two variables (quantity and price) shift in opposite directions. However, the model considers the costs and benefits from both an individual and social viewpoint. In fact, for society as a whole, the damage suffered by a subject could be offset by the benefit gained by another subject. For example, if the hydroelectric production is reduced, the plant owner suffers, but, in compensation, the energy could be produced by another plant that would have otherwise remained inactive.

Therefore, for society as a whole, the cost is not equal to the first subject's loss, but, instead, the difference between how much the first loses and how much the second gains. Reasoning thus, we can see in Table 1 how the effects of the dry climatic event, following the management strategy adopted, can be quantified in a total cost of 888 million euros. Yet, with a further breakdown, we can realize that, as a whole, agriculture comes out in front. Faced with a harvest loss of 749 million euros, the price increase verified results in other companies actually obtaining a benefit of 1.37 billion euros.

For the hydroelectric energy producers, there were no appreciable repercussions, as the possible reduction in productive capacity resulting from the earlier outflow was, fortunately, offset by a plentiful rainfall inflow in the following months. This restored the water body levels and, thus, the productive potential.

Instead, the consumers suffered from the impact of the higher prices resulting from a lower production, for a total loss of 1.5 billion euros.

Table 1 Net costs in managing the water stress on players involved

Players involved	€ Mill.
Farmers	**−628**
Production loss	749
Price increase	−1377
Electric energy producers	
Consumers	**1.516**
Loss in well-being—agricultural production	91
Price increase in agricultural products	1377
Loss in well-being—industrial energy use	22
Widespread loss in well-being	26
Total drought cost	**888**

Source Massarutto and de Carli (2009)

It is interesting at this point to analyse what could have occurred if different management scenarios had been adopted. The scenarios explored in the study forecasted:

- for agriculture, that it would not suffer withdrawal reductions, or that it could benefit from the extra quantities released from upstream. Alternatively, there could be a reallocation of the water for the different crops, favouring the high value-added ones, and changing the crop choices to crops requiring less water;
- for energy, where there is a deficit offset by a higher use of the thermal power station, or, alternatively, by a cut-off to users.

The result clearly shows that the first package of measures—neither reducing nor increasing the water supply—has a low impact. Only a limited number of companies could avoid the reduction in produced quantity, and, therefore, the reduction in damages would be negligible. Instead, a reallocation favouring high value-added crops and/or a change in production choices could significantly reduce the costs (up to 75 % less in the most ambitious of the scenarios). However, it should be noted that these scenarios are based on the effective possibility of water reallocation amongst the irrigated crops, which requires huge investments in the networks, thus resulting in a supply “on demand” (the networks would have to be converted into pump systems, and the fee structure changed to be able to use the immediate pricing as a means of reducing the consumption of the less productive uses).

The social costs of a power blackout would be potentially extremely high (0.67 billion euros only for cutting off industrial users with interruptible contracts), while the costs for substituting the usable sources are more limited, increasing in the short term the supply from the thermal power station or from abroad.

Very similar results to those obtained from the Po river basin study described above can be seen in the study on the effects of drought in another irrigated area, the Fruilian plains (Massarutto and Graffi 2012). Here, the model was created so as to specifically consider the possibility of redistributing the water supplied by the present management model (mainly a gravity-induced supply, and therefore, based on rigid rotation and limited possibility of transferring water from one company to another) and the costs of alternative options (investments to put pressure on the networks and so establish an on demand supply).

In the case of Friuli, the estimated critical frequency was 5 years. This means that if the time between the dry events was longer, then it would be better for the community to "run the risk" of choosing more profitable, but also more vulnerable, crops. If the events were more frequent, other strategies would be preferable (risk diversification through more balanced crop choices, investment in new irrigation methods, etc.).

Therefore, the 2003 event has been a lesson and left some quite clear policy implications.

The first is that, contrary to what could be feared, the impact on the agricultural activities as a whole is quite moderate and could be managed using mutual insurance tools (e.g. an eventual compensation for the farmers who lose their harvest borne by those who, instead, obtained a benefit).

The second is that, in the short term, the inflexibility of the system plays in favour of strategies aimed at minimizing the potential damage; however, once the emergency is overcome, it would be advisable to consider adopting medium to long-term projects aimed at reducing overall vulnerability. For example, this could be obtained through a more balanced crop selection. However, this is valid only if the frequency of water stress events exceeds a certain critical threshold.

The third is that the potentially serious effect in terms of social costs, namely a power blackout, could be averted if the system equipped itself with a reserve capacity, able to replace the hydroelectric supply, making use of other energy sources.

5 The Ecological Footprint of Water Consumption in Italy

Apart from the implications concerning the best strategies to face the future phenomena of water scarcity, the case analysed provides some useful recommendations, also from this book's analytical viewpoint.

As far as a water footprint approach is concerned, Italy is undoubtedly a country with a high internal[2] water footprint—water withdrawals are amongst the highest in the world but efficiency in use amongst the lowest. Italian agriculture, in particular, consumes an enormous quantity of water, and the more it uses, the less efficient its

[2]The internal water footprint is an indicator of the consumption of internal water resources in a delineated geographical area over a given period of time. It is different to the external water footprint which refers, instead, to the consumption of water resources originating from other countries (Hoekstra et al. 2011).

management model appears. With the right measures, water withdrawal could be considerably reduced as, for example, a great deal of water is presently used to irrigate very low value crops.

A misleading interpretation of these indicators is, however, to associate the use of water with "consumption" and, therefore, with environmental "dissipation". This approach is quite intuitively easy, but is not always correct.

The reduction in the water volume used can be or not be a desirable objective, but, in general, this is not so much due to the fact that we use a lot (in absolute values) but that we use it badly (i.e. using it for a limited social benefit). What should be considered is the degree of conflict amongst the competing uses—obviously also including the "environmental use", namely water for the production of ecosystem services—as well as the type of water to be used. In fact, the distinction between "green" and "blue" water allows for a much more precise evaluation of the impact of water resource use, as it takes into account the different cost-opportunities of the different virtual water "sources".

In short, the true "footprint" to be reduced is not so much the quantitative one (how much water we use), but rather the one that occurs due to a chaotic, uncoordinated and disorganized model of accessing a common good. If this was managed more effectively and efficiently, all the social demands, including those of the ecosystems, could be easily satisfied.

Italy, a water-rich country, especially in the north, has been, for some time, experiencing situations of water stress. These are due to the accompanying effects of a demand which, even though lower than in the past in absolute terms, is more inflexible because of the increased vulnerability of water-intensive economic activities, and of a supply which, due to climate changes and a stronger focus on ecological considerations compared to the past, has witnessed an overall reduction in usable resources and more frequent critical seasons.

The 2003 event resulted in social costs estimated at 1.5 billion euros only for the Po river basin. Costs that could have been much lower if the system had been better equipped and organized to tackle the situation, and if it had not had to, for the umpteenth time, face it as an emergency.

It is clear that Italy must rethink its water management model. I believe that the path to change lies mainly in searching for a way to reduce the present vulnerability, with more flexibility and ability to adapt, rather than in merely reducing the volume of water used.

Bibliography

Hoekstra, A. Y., Chapagain, A. K., Aldaya, M. M., & Mekonnen, M. M. (2011). *The water footprint assessment manual: Setting the global standard*. London: Earthscan.

IRSA-CNR. (1999). *Un futuro per l'acqua in Italia*, Quaderni di ricerca IRSA-CNR, n. 109, Rome.

Massarutto, A., & de Carli, A. (2009). I costi economici della siccità: il caso del Po. *Economia della fonti di energia e dell'ambiente*, *LII*(2), 123–143.

Massarutto, A., & Graffi, M. (2012). *Optimal strategies for managing drought vulnerability in water-rich contexts: Evidence from Fruili-Venezia Giulia*, DIES working paper in economics, University of Udine (to be published).
OECD. (2008). *Environmental outlook to 2030*. Paris: OECD.

From the BCFN's Double Pyramid to Virtual Water in the Production of Pasta Barilla

Luca Ruini, Laura Campra, Carlo Alberto Pratesi, Ludovica Principato, Massimo Marino and Sonia Pignatelli

Abstract We realize that water is a resource only when it becomes scarce. Until now, the issue seemed to interest only the least fortunate countries in the world, but this could all change: firstly because "high-quality" water—non-polluted freshwater—represents only a small part of the planet's reserves and secondly because of the increasing demand for water due to both the growing world population and more widespread wealth, which spurs more people in more countries to use (and waste) more water. Water use should be considered in both "real" terms (calculating the amount of water used for bathing, cooking, cleaning, etc.) and "virtual" terms (i.e., water footprint), estimating the total amount of water used in the entire life cycle of any product of service.

We realize that water is a resource only when it becomes scarce (Hanemann 2005). Until now, the issue seemed to interest only the least fortunate countries in the world, but this could all change: firstly because "high-quality" water—non-polluted freshwater—represents only a small part of the planet's reserves and secondly because of the increasing demand for water due to both the growing world population and more widespread wealth, which spurs more people in more countries to use (and waste) more water. Water use should be considered in both "real" terms (calculating the amount of water used for bathing, cooking, cleaning, etc.) and "virtual" terms (i.e., water footprint), estimating the total amount of water used in the entire life cycle of any product of service. It has been proven that agriculture is the phase in which the largest amount of water is used. By changing our diet—e.g., by eating more fruits, vegetables, and grains, while limiting animal protein—we could reduce our "virtual water" consumption, perhaps significantly (Hoekstra 2008).

L. Ruini (✉) · L. Campra
Barilla G. e R. Fratelli S.p.A, Parma, Italy
e-mail: Luca.Ruini@Barilla.com

C.A. Pratesi · L. Principato
Università Roma Tre, Roma, Italy

M. Marino · S. Pignatelli
Life Cycle Engineering, Turin, Italy

© Springer International Publishing Switzerland 2015
M. Antonelli and F. Greco (eds.), *The Water We Eat*,
Springer Water, DOI 10.1007/978-3-319-16393-2_15

Therefore, if demand rises and supplies decrease—also due to pollution and climate change—the value of water is bound to increase, and the current inequalities between water-rich and water-poor populations will generate further conflicts. We know how control over oil reserves has sparked interests and fueled dramatic feuds, and wars over water could be even worse. Because, after all, we can survive without oil but not without water. According to Tony Allan, the fact that the Middle East is able to import food from countries that do not suffer from water scarcity has avoided conflicts for many years. Thus, virtual water can bring balance to shortages in dry countries, and consequently foster peace and stability in areas that would otherwise be engaged in water wars.[1]

Only 2.5 % of the world's water resources are available for human consumption, and 85 % of that share is used in agriculture. This is the reason why we all need to make an effort to foster a more sensible use of water, both in agriculture and in our daily lives, by choosing a diet with a lower environmental impact.

Indeed, according to a recent study,[2] if we persevere in producing and consuming food in the same way we are doing now, in the long run, we will cause serious food crises in various regions of the world. In other words, our dietary mistakes can cause us health problems, but can also determine noteworthy impacts on the environment (International Water Management Institute 2007).

1 The Barilla Center for Food and Nutrition's Double Water Pyramid

Barilla Center FOR FOOD & NUTRITION

Spurred by this awareness of the impacts that food choices can have on the environment, in 2010, the Barilla Center for Food and Nutrition (BCFN) developed the Double Food–Environmental Pyramid model: a tool that compares the nutritional aspect of foods with their environmental impact.

The first pyramid represents the different food groups according to the principles of the traditional Mediterranean diet: At the base, there are fruits and vegetables, which are low in calories and high in nutrition (vitamins, minerals, and water) and

[1]Allan J.A. (2003), "Virtual water eliminates water wars? A case study from the Middle East", in Hoekstra A.Y. (editor), *Virtual water trade,* Delft, UNESCO-IHE, pp. 137–145. Allan J.A. (2011), "Virtual Water: Tackling the threat to our planet's more precious resource". London, New York: I.B Tauris.

[2]Molden D. (2007), *Water for Food, Water for Life: A Comprehensive Assessment of Water Management in Agriculture*, Earthscan.

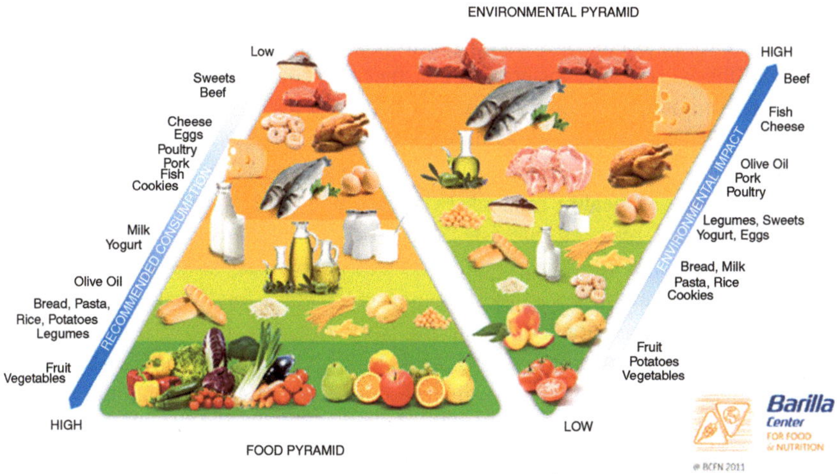

Fig. 1 The double food–environment pyramid. *Source* BCFN (2011a, b)

protective compounds, such as fibers. On the following levels of the pyramid, there are the other food groups with increasingly high calorie content, which are a staple of the North American diet although they should be eaten less often.

The environmental pyramid was created using data made available by scientific literature on the estimated environmental impact of each type of food, evaluating its whole life cycle assessment (LCA). This method evaluates a certain food's environmental impact by analyzing the entire supply chain, from raw matters to final product, including the management of any waste produced. The evaluation produces three quantitative indicators: ecological footprint (a measure of our Earth's ability to regenerate the resources employed), carbon footprint (the amount of greenhouse gas emissions produced), and water footprint (the amount of water resources used).

As shown in Fig. 1, the double pyramid points out that the foods having the highest impact on the environment are also the ones that we should eat less often (e.g., red meat), while the foods we should base our diet on (e.g., fruits and vegetables) have a lower environmental impact.

Although the BCFN's environmental pyramid was designed considering foods' ecological footprint, it could also take into consideration the water footprint.

Based on this indicator, a "water pyramid" could be placed next to the popular food pyramid, showing the relationship between environmental impact—in terms of water consumption—and recommended intake for each food group.

Box: The Barilla Center for Food and Nutrition

The BCFN analyzes data and suggests solutions with a multidisciplinary approach, with the goal of furthering knowledge about the global issues related to food and nutrition. Founded in 2009, the BCFN is a sounding board for the current needs of society and a hub of skills and experiences from around the world, thus fostering a constant and open dialog.

The complexity of the phenomena it investigates has led it to embrace a method that goes beyond the boundaries of single subjects. Thus, the issues it focuses on have been divided into four broad areas: food for sustainable growth; food and health; food for everyone; and food as culture.

The BCFN's research interests include science, the environment, culture, and economics. Within these fields, the BCFN conducts in-depth studies on relevant issues and offers suggestions on how to face food challenges of the future. Its activities are led by an Advisory Board made up of experts in complimentary fields, who suggest, analyze, and develop certain themes, and subsequently follow up with practical recommendations. The board currently includes, among others, Umberto Veronesi (oncologist), Gabriele Riccardi (nutritionist), Camillo Ricordi (immunologist), Claude Fischler (sociologist), Barbara Buchner (climate and environment expert), and Riccardo Valentini (agriculture, climate and environment expert).

The BCFN organizes events and presentations for public institutions and civil society organizations, such as the International Forum on Food and Nutrition, an important event for the leading figures in the field. The sixth edition ended in December 2014.

As there are no public data on the water footprint of fish, this food was not included in the graphical representation of the water pyramid. However, we can note it would have been important to consider both the amount of gray water used for fishing and industrial processing, and the amount used to grow food for aquaculture (fish farming). In any case, fish intake should be limited for reasons relating to the conservation of endangered species (yellowfin tuna, Atlantic bluefin tuna, swordfish, shark, etc.), regardless of virtual water consumption.

The outcome is an inverted pyramid in which different food groups are sorted by environmental impact in terms of water footprint: On top, there are the higher-impact foods and on the bottom the lower-impact ones.

In particular, we can note that—like in the ecological footprint model—red meat yields the highest water footprint, while fruits, vegetables, and grains have a considerably smaller effect.

By setting the food pyramid and the water pyramid next to each other (Fig. 2), we can see that most of the foods that nutritionists advise us to eat more often also have the smallest environmental impact in terms of water consumption. Vice versa, most of the foods we are supposed to limit also yield a high water footprint.

Therefore, if different food groups have different impacts on the world's water resources, what is the effect of our dietary habits on the environment?

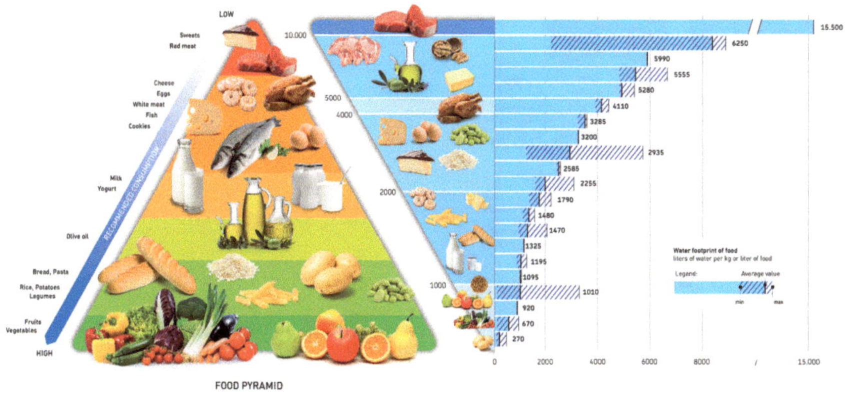

Fig. 2 The double food–water pyramid. *Source* BCFN (2012)

Box: The Importance of a "Water Economy"

Given the current demographic and economic trends, the BCFN has pinpointed the management of water resources as one of the world's greatest global challenges. Its position paper *Water Economy*, published in 2011, highlighted all the issues related to "high-quality" water, i.e., unpolluted freshwater, which represents only a very small part of the available water reserves and is a precious resource that must be managed with the utmost care. The world population growth and the level of wealth achieved in many countries drive people to use (and waste) more and more water, entailing the need to increase water resources.

With this increase in demand and decrease in availability, the economic value of water is bound to grow over time, causing imbalances that could lead to regional conflicts, perhaps even more serious and brutal than those for oil. Thus, we need to make a joint effort to promote a more sensible use of water, especially in agriculture and in our daily lives, choosing a sustainable food regimen, such as the traditional Mediterranean diet.

An individual uses an average of 2–5 L of water a day for drinking, while his or her daily consumption of virtual water embedded in food ranges from about 1500–2600 L for vegetarians to about 4000–5400 L for those following a meat-rich diet.

To better understand these differences, the BCFN detailed two different (but equally nutritionally balanced) daily menus, calculating their impact in terms of water consumption (as well as soil consumption and greenhouse gas emissions). The first menu is coherent with a diet rich in proteins from plant sources and low in animal fat, while the second one includes red meat, albeit in limited quantities. Comparing the water footprint of the two menus clearly shows that even a limited

Fig. 3 A sample vegetarian menu and one including one meat dish, with their respective environmental impacts and water footprints. *Source Doppia piramide*, BCFN (2012)

intake of animal products, like dairy and meat, causes the consumption of water resources to nearly treble.

Indeed, milk and meat have a greater virtual water content than fruits and vegetables, due to the high consumption of agricultural products used to feed farm animals with the goal of transforming them into sources of food. Animals raised on pasture do not pose the same problems, as raising them does not require large amounts of high-energy foods such as corn.

These examples clearly show that changes in individual eating habits can have a very significant impact on the availability of water resources. Suffice it to say that if everyone on the planet adopted the average Western diet, which is characterized by high intakes of meat, we would need 75 % more water than what is currently used in the world to produce food.[3]

While fifty years ago the global population's size and lifestyle determined lower levels of water consumption, today the competition for increasingly scarce water resources has become much more intense: Many basins are now unable to satisfy the local demand for water, and others have been completely drained. In perspective, the water shortage will be a constraint on food production for millions of people, forcing them to import food from other countries (Fig. 3).

[3]Zimmer D. and Renault D. (2003), "Virtual Water in Food Production and Global Trade: Review of Methodological Issues and Preliminary Results", Hoekstra A.Y. (editor), *Virtual Water Trade: Proceedings of the International Expert Meeting on Virtual Water Trade,* Value of Water Research Report Series n. 12, Delft, UNESCO-IHE, Institute for Water Education.

The situation clearly requires a drastic turnaround, in which the responsibility of institutions and businesses (both in agriculture and the industry) must be bolstered by citizens' personal commitment to contribute to the progressive reduction of consumption and waste.

This could be achieved by simply improving our eating habits, choosing diets that are healthier for us and help significantly reduce our water footprint. In order to educate people on this issue, the Double Pyramid model was selected for a presentation at the Village of Solutions during the Sixth World Water Forum (see "The Double Water Pyramid at the Village of Solutions during the 2012 World Water Forum").

But does eating in a healthy and sustainable way cost more?

According to the results published by the BCFN in the third edition of the *Double Pyramid* report, choosing a diet that is nutritionally balanced and environmentally sustainable does not necessarily entail an additional expense for the consumer: Indeed, in some cases, there might be added savings.

In Italy, for example, the information provided by the Ministry of Economic Development's Price Observatory has allowed the BCFN to determine that the weekly cost of food for someone cutting back on meat in favor of fruit, vegetables, and grains can decrease by as much as 10 %. However, this is not true for all countries; for example, in the United States, a diet rich in animal protein is cheaper, and the data regarding France and the United Kingdom are conflicting, with some studies coming up with higher and others with lower estimates.

Box: The Double Water Pyramid at the Village of Solutions During the 2012 World Water Forum

In March 2012, the sixth World Water Forum—the largest global conference on water promoted by the World Water Council and the International Forum Committee—was organized in Marseille. The event developed through the work of various committees, which addressed and proposed solutions for 12 global priorities, including the need to ensure the world population's health and well-being and the implementation of sustainable economic development that allows the planet to remain "blue." The aim was to foster dialog on these issues and to suggest solutions that may be considered by international policy makers.

For the debate on water economy and the management of water resources, the World Water Forum Committee selected the BCFN's Double Food-Environment Pyramid and Water Pyramid as one of the most effective ideas among the many proposals received; the models were showcased in a dedicated space called "Village of Solutions." The concept underlying the double pyramids was also presented in the Agora of the Village, gaining great response from both the general public and experts (World Water Forum 2012).

1.1 Barilla's Efforts for Its Products' Sustainability Certification, and for the Reduction of Water Consumption in Its Facilities

So far, we have seen the activities of the Barilla Center for Food and Nutrition as a think tank and independent research center. In this second part of the chapter, we will analyze the steps taken by Barilla itself to streamline and reduce water consumption at the company's pasta factories and bakeries, and along the products' entire life cycle, also in order to reliably estimate their water footprint.

Since 2000, the company has been conducting LCAs on its products, with the aim of improving their environmental performance year after year. These analyses follow objective procedures for the evaluation of energy and environmental impacts related to a process or activity, carried out by identifying the energy and materials used and any waste released into the environment. The assessment encompasses the entire life cycle of the process or activity, including extraction and processing of raw materials, manufacturing, transportation, distribution, use, reuse, recycling, and final disposal.

Since 2008, Barilla has begun to evaluate its products' water footprint, based on LCAs and on the calculation protocol developed by the Water Footprint Network.

In 2012, life cycle analyses covered over 53 % of Barilla's production (in terms of volume).

In February 2011, Barilla was the first private company in the food sector to certify a system for the calculation of products' environmental impact according to the guidelines provided by the Environmental Product Declaration (EPD®) International System.

The EPD® is a public document reporting a product's environmental impact over its whole life cycle, from the production of raw materials up to distribution and disposal, if applicable, once the product has exhausted its function.

The statement provides data on key environmental impacts, such as the carbon or water footprint, as well as qualitative information about the processes and policies implemented by the organization that produces and markets the product.

Barilla has decided to adopt the EPD® as a tool for the calculation and communication of its products' environmental impacts because the validation by an external subject ensures that the information is fair, accurate, and reproducible. The EPD® is also the only system that fully complies with ISO standards, which require public assessment of calculation methods.

Barilla believes there are several reasons why it is important to know the impact of its products: first of all, to identify possible improvements that may be implemented along the supply chain; secondly, so that all levels of the organization are on the same page when discussing environmental matters; and finally, to have the most reliable and solid information to communicate outside of the company.

Box: The Environmental Product Declaration of Barilla's Durum Wheat Flour Pasta

The factory and process

Durum wheat flour pasta is made "by extrusion, rolling and drying dough that has been prepared with durum wheat flour and water." The Barilla products that fall within the scope of this EPD® are classic pasta shapes (penne, rigatoni, spaghetti, etc.); "Piccolini" (miniatures of the classic shapes); "Specialità" (special shapes like barbine, castle, and bowties); and "Regionali" (traditional shapes from different Italian regions, such as gnocchi, orecchiette, and reginette). These products differ only in shape, as they are all made using solely water and durum wheat flour.

Every year, the Barilla group produces approximately 1,000,000 metric tons of pasta in the eight factories it owns in five countries (Italy, Greece, Turkey, United States, and Mexico).

The product

The results presented here regard Barilla durum wheat pasta manufactured and consumed in Italy, packed in cardboard boxes.

The study was carried out according to EPD® rules for the product category "Code CPC 2371—Uncooked pasta, not stuffed or otherwise prepared."

The assessment considered all the following stages of production:

1. Wheat cultivation;
2. Flour production;
3. Pasta production;
4. Raw materials and products transport to distribution platforms.

Subsequent steps, related to cooking the pasta and disposing of its primary packaging, are closely connected to consumer behavior and therefore subject to greater variability. Thus, the impacts were estimated assuming that the amount of water used for cooking is exactly as recommended by Barilla and that the cardboard box is disposed of in average conditions.

The assessed system is broken down in the figure below, in which we can identify three different phases.

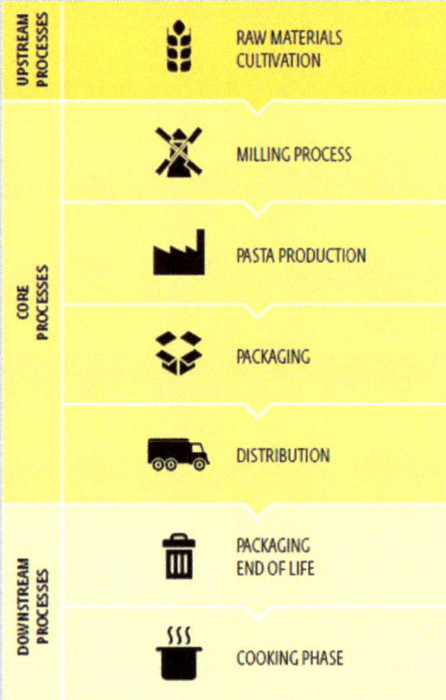

1. Raw materials (upstream processes), including the following:

   the cultivation of wheat, with specific information about the areas where wheat used by Barilla to make pasta is produced.

2. Pasta production (core processes), including the following:

   the production of flour by milling durum wheat, which takes place at facilities owned either by Barilla or by accredited suppliers. As regards this report, we considered only proprietor mills, which contribute 70 % of production. We assume that the impacts generated by the remaining 30 % provided by other manufacturers are equal to proprietor mills, since the production system and level of efficiency are the same;

   activities and processes for the production of Barilla pasta in Italian factories (Pedrignano, Foggia, and Marcianise);

   the processing of materials for packaging, mainly cardboard for primary packaging. The environmental performance associated with the production of packaging was evaluated per kilo of product, considering the heaviest pasta shape (all other shapes will have a relatively smaller impact in this phase). We used primary data (provided by the unit in charge of packaging design) to determine the amounts produced, and secondary data (from the Ecoinvent database) for the environmental aspects related to packaging production;

the environmental performances associated with distribution were evaluated by making specific assumptions for each production area. We used primary data about the distance covered by road, rail, and sea, as well as secondary data (from the Ecoinvent database) as regards the means of transportation.

3. The subsequent stages (cooking and packaging disposal) depend on consumers' behavior and are, therefore, beyond the scope of the assessment (downstream processes); we made assumptions, and based on those assumptions, we estimated the impacts associated with these activities.

   The results show that the agricultural phase of growing wheat and cooking have the greatest impact on the environment in terms of ecological, carbon, and water footprint.

The results of the LCA for Barilla pasta manufactured in Italy for the Italian market

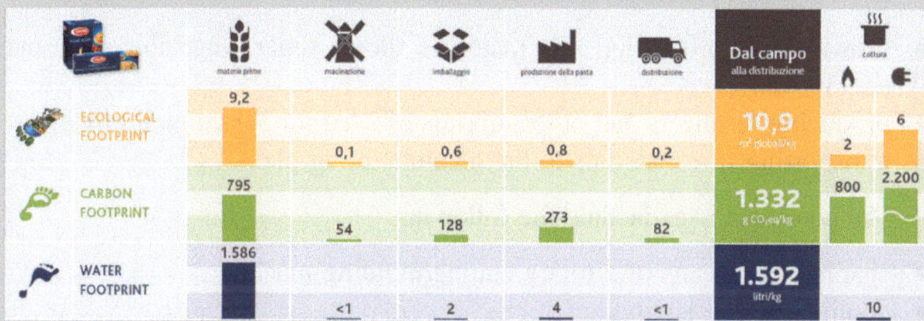

Source Barilla (2011a, b).

Box: The Environmental Product Declaration of Mulino Bianco-Barilla's Tarallucci

The Mulino Bianco brand was founded in 1975 and offers a variety of simple and wholesome baked goods. Tarallucci are one of the historic Mulino Bianco biscuits.

The factory and process

Tarallucci are produced in two Italian factories in Italy (Castiglione delle Stiviere, in the province of Mantova, and Melfi, in the province of Potenza). The production process includes the following stages: preparation of the dough, forming, cooking, cooling, and packaging. Packs of Tarallucci are available in two sizes: 400 and 800 g. In the analysis, we considered the 400 g format, since it uses a relatively larger quantity of packaging material (per kilo of product).

The environmental performance of Tarallucci was evaluated through the LCA method, starting from the production of raw materials and including every step until delivery to the main distribution platforms.

The study was carried out according to EPD® rules for the product category "Code CPC 2349—Bread and other bakers' wares."

All results and data refer to a functional unit of 1 k of finished, ready-to-use product.

The analysis includes all the processes that are part of the system subject to assessment, which were grouped into three phases, in compliance with the EPD® system's requirements.

The upstream processes include the following:

- growing and processing raw materials (flour, sugar, vegetable oils, and eggs);
- producing fertilizers and other substances used for agricultural processes;
- transporting raw materials to the manufacturing plant.

The core processes include the following:

- shaping, by means of a rotary cutting/molding machine;
- cooling and packaging;
- transporting products to distribution platforms.

The downstream processes include general information on end-of-life packaging. The evaluation of the environmental aspects linked to these processes is beyond the scope of the system.

The results show that even for Tarallucci the phase that has the greatest environmental impact by far is the one related to agriculture.

The results of the LCA for Mulino Bianco Tarallucci

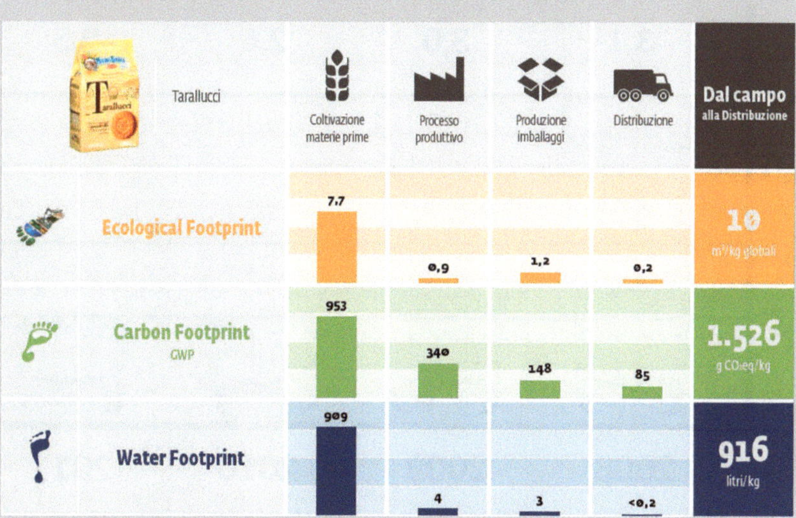

Source Barilla-Mulino Bianco (2011).

In August 2012, 21 environmental product declarations regarding Barilla pasta and bakery products were published on the www.environdec.org web site. In particular, the company published the EPD® for Barilla durum wheat pasta, four Wasa products, five Pavesi products, and eleven Mulino Bianco products.

As regards energy-saving efforts in factories and bakeries, it should be noted that 72 % of Barilla's products are placed on the market by facilities that are ISO 14001 certified, and thus comply with an international standard that ensures the proper management of all environmental aspects.

Programs and activities for the reduction of water consumption have been implemented for years, with significant results. Indeed, in 2011, Barilla factories consumed about 2.5 million cubic meters of water, saving about 800,000 m^3 compared to 2008 (approximately 19 %). Comparing 2011 with 2006, the percentage of water saved rises to 30 %. Furthermore, the new plant in Rubbiano uses 47 % less water compared to the average of factories that make pasta sauce (see "Saving water in Barilla's new pasta sauce factory in Rubbiano").

Let us not forget that water, in addition to being used as a raw material in some recipes, is also used for different purposes, such as washing and/or cooling off equipment, activating fire systems, irrigating green areas, allowing employees to wash, and preparing food and beverages for the cafeteria (Fig. 4).

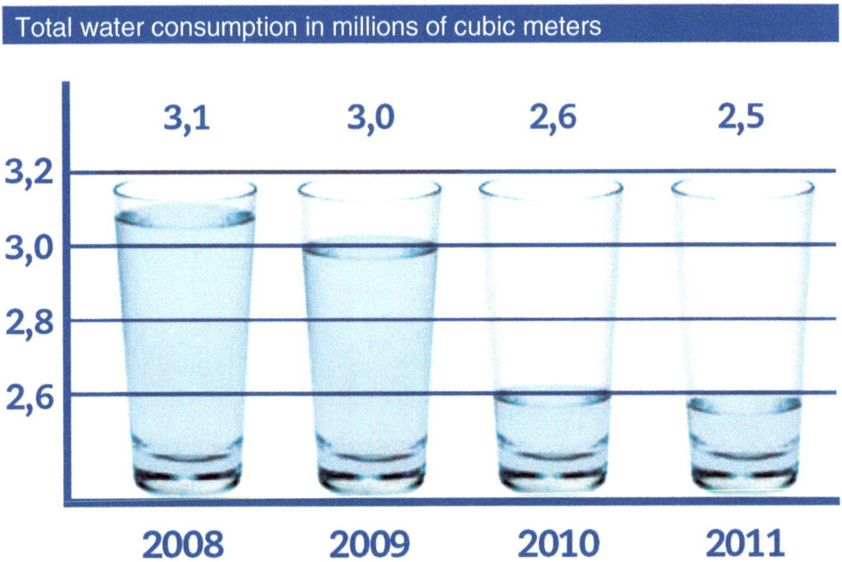

Fig. 4 Total water consumption in Barilla factories, in millions of cubic meters. *Source Barilla Sustainability Report*, 2012

In 2011, the pasta factories that recorded the greatest reduction in water consumption (per unit of finished product) were the ones in Caserta and Pedrignano in Italy, and the one in Ames in the United States.

The bakeries in Novara, Valenciennes, Castiglione, and Plain de l'Ain were able save at least 10 %.

The following activities contributed to the reduction of water consumption in Barilla factories:

- Eliminating the cooling systems that did not recycle water;
- Optimizing cooling tower management;
- Installing water flow regulators;
- Water recovery for the cooling towers in the plant in Cremona.

In addition, 60 % of Barilla manufacturing plants are equipped with a wastewater treatment system, which considerably reduces the impact of wastewater before disposal. In 2011, approximately 800,000 m^3 of water were discharged in the public sewerages (just over 80 %), while the rest was drained directly in surface water.

Box: Saving Water in Barilla's New Pasta Sauce Factory in Rubbiano

In October 2012, Barilla opened a new factory for the production of pasta sauces in Rubbiano di Solignano, in the province of Parma.

In the face of the current difficult and unstable economy, Barilla demonstrated its commitment to the Italian productive system with the creation of 120 jobs at the new factory, which will have a productive capacity of 60,000 metric tons per year.

Rubbiano will produce 160 metric tons of tomato-based sauces and 75 metric tons of pesto sauces, which will be exported all over the world. The plant was designed according to the highest technological standards and will be one of the most energy- and water-efficient factories for the manufacturing of pasta sauce in the world. In line with Barilla's commitment for environmental sustainability, it will produce 32 % less carbon dioxide and consume as much as 47 % less water compared to the average of pasta sauce factories.

A water footprint can be calculated for any product or activity, as well as for any specific group of consumers (an individual, a family, the residents in a city, or an entire nation) or manufacturers (private companies, public organizations, entire economic sectors (Hoekstra and Hung 2002).

The global water footprint is approximately 7.45 trillion cubic meters of freshwater per year, equal to 1240 m^3 per capita per year, i.e., more than twice the annual flow rate of the Mississippi river.[4]

Considering the water footprint in absolute terms, the country that consumes the largest volume of water is India (990 billion cubic meters), followed by China (880) and the United States (700).

[4]Barilla Center for Food and Nutrition (2011), *Water Economy*, available to view or download from Freebook Ambiente (freebook.edizioniambiente.it/libro/55/Water_Economy), Chap. 4.1 (in Italian).

However, if we take into account per capita values, the United States ranks first, with an average water footprint of approximately 2480 m^3 per year, followed by Italy (2230) and Thailand (2220). The differences between countries are dependent on a range of factors. The four main ones are the volume and pattern of consumption, climate, and farming practices.

Since raw materials, goods, and services all have a certain amount of virtual water embedded in them, trade between countries entails a transfer of virtual water flows (virtual water trade). Indeed, the water footprint or virtual water embedded in a product (whether it is a physical good or an intangible service) is determined by the total volume of freshwater consumed to produce it, considering all the various stages of the production chain. Water footprint can be subdivided into two elements: internal (i.e., domestic consumption) and external water footprint (the consumption of virtual water coming from other countries (Hoekstra 2008; Hoekstra et al. 2011).

Europe is a net importer of virtual water, and its water security is highly dependent on external resources. Water globalization seems to involve both opportunities and risks, as the degree of interdependence between countries in the exchange of virtual water is destined to grow, due to the ongoing process of international trade liberalization.

One of the main opportunities in this context is represented by the fact that virtual water can be considered as an alternative source of water, allowing a country to preserve its local resources.

Barilla published its first report on water during the World Water Week, held in Stockholm in August 2012. The report highlights the water footprint of the pasta manufactured by the company and assesses the amount of virtual water involved in the wheat and pasta trade among the nations in which Barilla operates.

The water footprint of Barilla pasta ranges from only 1350 L/kg, when produced in Italy, to over 2850 L when produced in Turkey or the United States. Therefore, it positions itself at the lower end of the pyramid of water, just above cereals and bread, but abundantly below rice and beans.

The water footprint of pasta comes almost entirely from the water used in the cultivation of wheat in different countries. Barilla buys durum wheat not only in Italy but also in other countries in Europe (France, Greece, and Spain) and outside of the continent (Canada, United States, Australia, and Mexico).

The water footprint of durum wheat cultivated in different geographical areas varies depending on the availability of water, climate conditions, and crop yields. The value ranges from 1000 L/kg in northern Italy and France to 2000 L/kg in Australia and Turkey.

In 2011, the total water footprint of durum wheat used by Barilla was about 2,000 million cubic meters, 5 % less than in previous years.

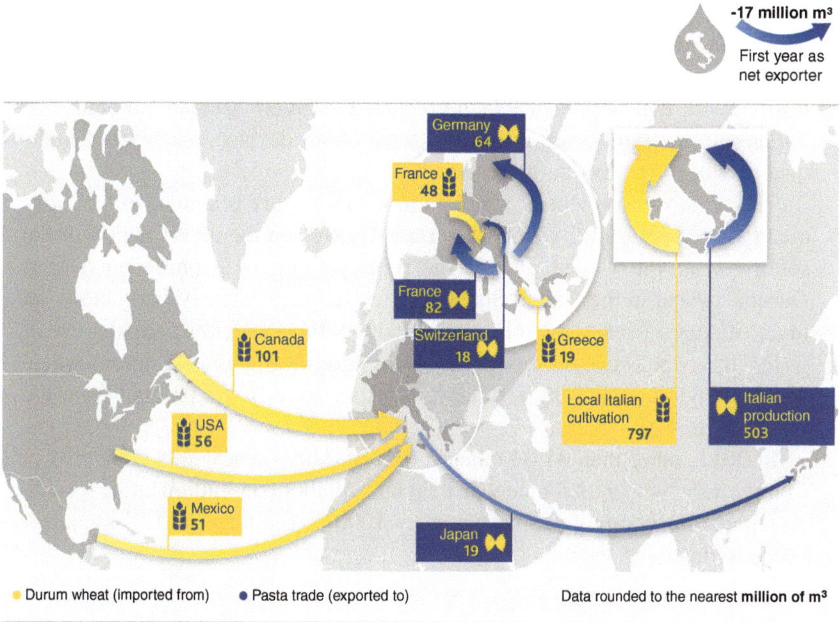

Fig. 5 Pasta and durum wheat, Barilla Italia 2011. *Source Water Report Barilla*, 2012

1.3 Virtual Water Flows for Barilla Pasta and Durum Wheat

The *Water Report* details virtual water flows from the last three years, from 2009 to 2011, for all the countries in which there is a Barilla factory.

Figure 5 summarizes the virtual water flows related to Barilla pasta and durum wheat in Italy. In this country, wheat production does not use "blue water"—the blue water footprint is the volume of water that is used (for irrigation, in the case of wheat) and not returned to its source downstream in the production process. Furthermore, the amount of imported blue water is small compared to the total volume of virtual water inherent in the importation of wheat (blue water represented about 9.9 % of the total in 2009 and 12.8 % in 2011). The amount of blue water in the trade of pasta is even lower, and amounts to about 5 % of total water flows per year in the period considered.

As regards the use of green water (the green water footprint is the volume of rainwater used by crops during the growth phase for evapotranspiration), importations from abroad represent a percentage ranging from 32.5 % in 2009 to 35.4 % in 2010 and 23.4 % in 2011. The decline in imports over the years is at least partly due to the introduction and development of Aureo durum wheat in Italy (see "The Aureo wheat project: 35 million cubic meters of blue water saved"). Saving blue water is very important because it decreases water consumption directly. In this case, indeed, moving wheat fields to a different area (from the desert region of Colorado in the United States to southern Italy) cut the need for irrigation water.

Box: The Aureo Wheat Project: 35 Million Cubic Meters of Blue Water Saved

Barilla, with the support of Italian breeder, Produttori Sementi Bologna, has initiated a program to use traditional methods to develop a variety of high-quality wheat that can be grown in Italy.

This led to Aureo wheat, which in terms of quality is similar to the high-quality Desert Durum® variety that Barilla grows in the desert region in the southwest of the United States. The project's aim is to phase out Desert Durum® wheat in the United States—which requires extensive irrigation—in favor of Aureo wheat grown in central Italy, with considerably smaller use of water; this would decrease both the water footprint of Barilla durum wheat and the environmental impact due to transportation of the product from the United States.

In 2011, more than 41,000 metric tons of Aureo wheat was cultivated in Central Italy, with a total blue water savings of about 35 million cubic meters. At regards transportation, approximately 1000 metric tons of CO_2 equivalent was saved.

Northern and Central Italy contribute in a similar way, respectively, with 35.4 and 38.7 % in 2010, while Southern Italy (mainly Puglia and Sicily) has a slightly lower share of (25.9 %).

The rest of the virtual water embedded in durum wheat comes from other countries, most importantly the United States, Canada, Mexico, and France.

Regarding virtual water exportations, 63.2 % of the pasta manufactured in Italy is destined to the domestic market, and the main countries in which Barilla exports are France (28.0 %), Germany (21.7 %), Japan (6.5 %), and Switzerland (6.2 %).

Moreover, approximately 62 % of total virtual water embedded in durum wheat used by Barilla for the production of pasta comes from local crops.

Figure 6 summarizes the incoming flows of water embedded in durum wheat and pasta used for Barilla's production in 2011. Flows labeled as "input" represent virtual water entering Italy through the wheat imported from abroad to produce Barilla pasta in Italy.

Flows labeled "output" (regarding pasta) represent virtual water leaving the country through the pasta that is exported to the countries listed. Unlike previous years, in 2011, the export of virtual water in Barilla pasta produced in Italy exceeded the amount of imported water linked to the production of durum wheat.

Should the scenario of virtual water trade continue along this trend in the coming years, a well-balanced water saving policy might be implemented. Therefore, pasta manufacturers and the sector revolving around them could make a significant contribution to virtual water trade, also considering the fact that Italy is a major importer of virtual water.

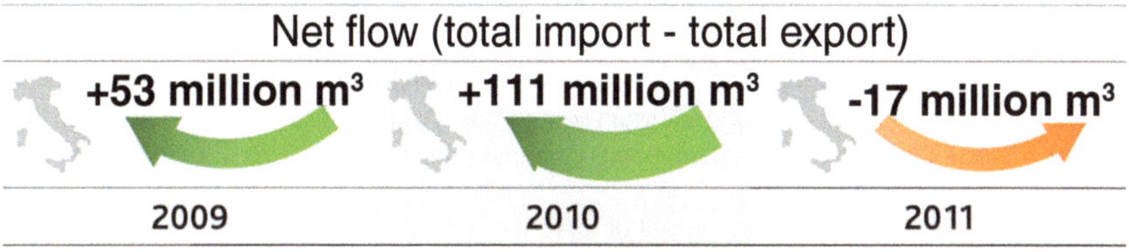

			2009 INPUT [.000 m³]	2010 INPUT [.000 m³]	2011 INPUT [.000 m³]
Durum wheat	Local production		**707.141**	**697.748**	**796.757**
	imported from	France	15.578	28.046	47.643
		Greece	-	27.558	19.452
		Australia	120.254	66.700	-
		Mexico	41.687	64.548	50.670
		USA	138.562	101.999	55.772
		Canada	34.585	98.340	101.449
		other	14.069	33.859	-
		Total	364.735	421.051	274.986
	Total quantity		**1.071.876**	**1.118.798**	**1.071.743**

			2009 OUTPUT [.000 m³]	2010 OUTPUT [.000 m³]	2011 OUTPUT [.000 m³]
Pasta	Local production		**536.831**	**534.353**	**502.887**
	exported to	France	87.513	87.109	81.979
		Germany	68.063	67.748	63.759
		Japan	20.497	20.403	19.201
		Switzerland	19.299	19.210	18.079
		other	116.578	116.040	109.207
		Total	**311.950**	**310.510**	**292.225**
	Total quantity		**848.781**	**844.862**	**795.112**

Fig. 6 Virtual water imported with durum wheat (*top*) and with pasta (*bottom*). *Source Water Report Barilla*, 2012

References

Allan, J. A. (2011). *Virtual Water: Tackling the threat to our planet's more precious resource*. London, New York: I.B Tauris.

Barilla. (2011a). 2010 Sustainability Report.

Barilla. (2011b). Environmental Product Declaration of made-in-Italy durum wheat dried pasta in paperboard box, Revision 2.1 (valid for 3 years after approval, registration number S-P-00217, date of approval March 10, 2011).

Barilla. (2012a). *Good for you, Sustainable for the planet… in other words, our way of doing business*. 2011 Sustainability Report.

Barilla. (2012b). *Water Report 2012*. Stockholm water week draft, version 1.

Barilla—Mulino Bianco (2011), Environmental Product Declaration of Tarallucci biscuits, Revision 1 of March 10, 2011, Certification n. S-P-00226.

Barilla Center for Food and Nutrition. (2011a). *Water economy*, Parma.

Barilla Center for Food and Nutrition. (2011b). *Double pyramid: Healthy diet for all and environmentally sustainable*. Parma.

Barilla Center for Food and Nutrition. (2012). *Doppia piramide: Favorire scelte alimentari consapevoli*. Parma.
Hanemann, M. (2005). *The value of water*. Berkeley: University of California.
Hoekstra A. Y. & Hung P. Q. (2002). Virtual water trade: A quantification of virtual water flows between nations in relation to international crop trade. *Value of Water Research Report Series*, vol. 11. Delft: UNESCO-IHE.
Hoekstra A. Y. (2008). The water footprint of food. *Water for food. The Swedish Research Council for Environment*. (pp. 49–60).
Hoekstra, A. Y., Chapagain, A. K., Aldaya, M. M., & Mekonnen, M. M. (2011). *The water footprint assessment manual: Setting the global standard*. London: Earthscan.
International Water Management Institute. (2007). *Water for food, water for life: A comprehensive assessment of water management in agriculture—summary book*. Colombo, London: Earthscan.
Mekonnen M. M. & Hoekstra A. Y. (2010). The green, blue and grey water footprint of crops and derived crop products. *Value of water research report series*, vol. 47. Delft: UNESCO-IHE Institute for Water Education.
Molden, D. (2007). *Water for food, water for life: A comprehensive assessment of water management in agriculture*. London: Earthscan.
Renault D. (2002). Value of virtual water in food: Principles and virtues. In A. Y. Hoekstra (Ed) *Proceedings of the Expert Meeting*, UNESCO-IHE, 12–13 Dec 2002.
Renault D. & Wallender W. W. (2000). Nutritional water productivity and diets: From 'crop per drop' towards 'nutrition per drop'. *Agricultural Water Management*, *XLV*, 275–296.
World Water Forum. (2012). *The "Double Water Pyramid" to promote sustainable diets*. Village of Solutions, Knowledge Pavilion. Solution for water n. 2348.
Zimmer D. & Renault D. (2003). Virtual water in food production and global trade: Review of methodological issues and preliminary results. In A. Y Hoekstra (Ed.), *Virtual water trade: Proceedings of the International Expert Meeting on Virtual Water Trade*. Value of Water Research Report Series n. 12. Delft: UNESCO-IHE, Institute for Water Education.

Part IV
Water in Food

The Virtual Water in a Bottle of Wine

Lucrezia Lamastra

Every form of life on the Earth depends on water, and it is in water that billions of years ago the first life forms appeared. Also today, the almost 9 million living species found on our planet base their existence on water, a resource which is, therefore, not only essential but also very precious. Despite it being a renewable resource, it still, however, remains limited and vulnerable. Even if our planet viewed from afar appears as a prevalently blue sphere, with 71 % of the surface covered by water, we know very well that not all this water is actually available to humans. First of all, 97 % of the water is salt water found in the seas and oceans and only 3 % is freshwater, of which, however, most (68.6 %) is locked up in ice and glaciers, 30.1 % in groundwater and 1.3 % in surface water. The liquid water on the land surface is mainly found in the large lake basins, such as the North American Great Lakes or Lake Baikal in Russia, which contain 20.1 %, equal to 0.26 % of total freshwater, and in the swamps which make up 2.53 % (0.03 % of total freshwater). The atmosphere contains 0.04 % of total freshwater in the form of water vapour and the land 0.05 %, while the river systems contain a relatively low portion (0.006 %) (Fig. 1). Moreover, the geographic distribution of water is not homogenous—Brazil has 15 % of the global reserves and 64.4 % of the total water found on the Earth is found in only 13 countries (http://ga.water.usgs.gov/edu/earthwherewater.html; Shiklomanov 1999). For this and for reasons of economic inequality, despite only 54 % of the world's freshwater reserves presently used being accessible, a billion people do not have access to drinking water and 2 billion people do not have sufficient water for hygiene-sanitary services (Prüss-Üstün 2008; IWMI 2007).

The term water stress was coined by the World Resources Institute (WRI) to indicate when, in a delineated zone, there is not sufficient water to meet agricultural, industrial or domestic needs (Revenga et al. 2000). Therefore, we say a zone is suffering water stress when the annual freshwater availability is less than 1700 m^3

L. Lamastra (✉)
Institute of Agrarian and Environmental Chemistry,
University "Cattolica Del Sacro Cuore", Milano, Italy
e-mail: Lucrezia.Lamastra@unicatt.it

© Springer International Publishing Switzerland 2015

M. Antonelli and F. Greco (eds.), *The Water We Eat*,
Springer Water, DOI 10.1007/978-3-319-16393-2_16

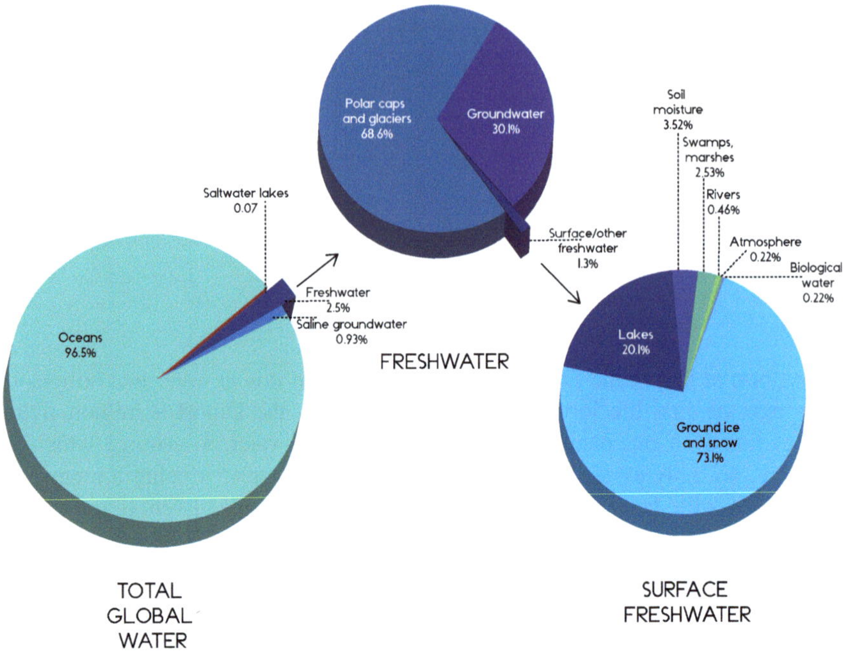

Fig. 1 The distribution of water on the planet. *Source* Shiklomanov (1993)

per year per capita. Instead, we say there is a chronic water scarcity when the availability falls between 500 and 1000 m^3 a year per capita, while absolute water scarcity is when the quantity falls below 500 m^3 a year per capita. Corresponding to these levels, there are profound economic impacts on development and serious risks to human health (Rijsberman 2006; Falkenmark 1992). The present estimates indicate that, by 2025, water stress will be a reality for almost half of the world's population. This, in turn, will lead to the price of water, which reflects the supply scarcity and competition, continuing to increase and thus changing the water allocation to the different production sectors and user groups (Rosegrant 2002; UNDP 2006). Added to the problems of local and regional water shortages, there is the problem of pollution which renders enormous volumes of water unsuitable for civil and non-civil use. Pollution threatens the quality of water resources and is also linked to demographic growth and the access to the market of large swathes of populations previously excluded from mass consumption, which has resulted in an increase in production and in the management of its waste. For example, it is estimated that 2 million tonnes of waste generated from human activities are returned to water courses with direct consequences on their quality. In fact, in developing countries, 90 % of effluent water and 70 % of industrial water are released into water bodies without being subjected to any kind of treatment, polluting the freshwater resources available for humans. Moreover, the water cycle is also affected by the climate change, the acidification of the oceans, the melting of the ice caps, the increase in the average sea level, the shifting of tropical storms towards the poles, with significant effects on the winds, rainfall and temperatures, and an increase in the frequency of "extreme" phenomena, such as flooding and heat waves (UNESCO 2009).

1 Impacts and Footprints: The Sign on Water

As far as the allocation of water resources is concerned, it should be remembered that most freshwater is employed by agriculture (more than 70 %), 23 % by industry and 7 % by domestic use (FAO 2012).

These factors have led academics in the environmental sciences to look for a way to express the impact on the water resources of each production process and consumer good with the objective to guide and urge for the most sustainable resource use possible. Consequently, in the last ten years, the calculation of the water footprint has spread. The term "footprint", common to carbon footprint and ecological footprint, sets a quantitative measurement of the appropriation of natural resources by humans (Galli et al. 2012). Consequently, as for other footprints, the water footprint is also being increasingly applied in companies and is receiving more and more attention from consumers, setting underway an awareness of *how* and *where* water is used.

Often the terms "water footprint" and "virtual water" are used synonymously. However, some distinctions between the two should be made. Both express the volume of freshwater "contained" in a product, not really but virtually, that is, all the water used and polluted throughout the production process. However, the water footprint distinguishes itself from the virtual water content as it refines the concept —it expresses and distinguishes the different types of water used and it is spatio-temporal explicit. Therefore, it tells us what type of water was used, and how it was used, with a value changing over time and depending on the production site (Hoekstra et al. 2011). Moreover, while the concept of the virtual water content may be easily reduced to a product, it is difficult to correlate to a person or a group of people (nobody could easily understand the statement "the virtual water content of the average Italian is more than 2 million litres" (Mekonnen and Hoekstra 2011). The footprint, instead, is a concept which can be easily used also to refer to the impact on the water resources relevant to a consumer as much as to a group of people (organisation, city, region, nation, humanity).

In calculating the water footprint, three different typologies of water (green, blue and grey) have been individuated, which define the nature of the water used (Fig. 2). Green water is the volume of rainwater used by crops through evapo-transpiration and is extremely important for agricultural products. Blue water is the freshwater withdrawn from a water basin which is not then reintroduced into that basin, or it may return there but at different times. Finally, there is grey water, a new

Fig. 2 The water footprint: the 3 colours of virtual water. *Source* Hoekstra et al. (2011)

and intuitive way to describe water body pollution in terms of volume, and therefore, this quantity can be added to the previous other two to obtain a total index. Grey water expresses the "imaginary" water volume required to dilute the pollution possibly produced below established legal and/or ecotoxicological end points. It is not actually a quantity of water used, but is the quantity of water which would be necessary to cancel the pollution resulting from the process being analysed (Hoekstra et al. 2011).

Therefore, the water footprint does not represent the water actually contained in a product but how much is required to produce that product. It is an indicator of the water volumes used in the production process, considering as well the water required to neutralise the resulting eventual impacts (if we think, for example, that a 0.33 l can of Coca Cola has a water footprint of 35 l, it is quite clear that this volume does not exclusively reflect its actual content; Coca Cola Europe 2011).

Consequently, the theme of the virtual water concept brings to the fore the water consumption and pollution which can be verified throughout the entire supply chain, highlighting the volumes actually contained in the product though usually "hidden". It is because they are "hidden" that this consumption is often ignored and even today, there is a lack of awareness of how much production and supply chain aspects can affect the volumes of water consumed, on the spatiotemporal distribution of its use and on the resulting types of pollution. Very often, we could almost say always, these impacts on the water resources which occur along the supply chain greatly exceed the volumes regarding the product's actual water content and its direct water consumption (Molden 2010).

2 From the Field to the Table: The Virtual Water of Agri-Food Products

Over the last years, the growing interest in the question of water has led companies, institutions and individuals to look more carefully at the issue of virtual water and the water footprint and to tackle the calculation of its relative volumes. All the studies carried out have highlighted that the virtual water of agricultural products and their derivatives always corresponds to high volumes, and for the product derivates, the most important phase is always the agricultural one. We have already mentioned that 70 % of freshwater withdrawal is due to water use in agriculture, especially regarding irrigation practices which are increasingly spreading worldwide, and which are a factor of climate change and production intensification. Mekonnen and Hoekstra in the report, "National Water Footprint Accounts: the green, blue and grey footprint of production and consumption", published in 2011, show that agriculture is responsible for about 92 % of the global water footprint (9087 Gm^3/year for the period 1996–2005; Mekonnen and Hoekstra 2011). However, it is necessary to make a distinction here. Clearly, each crop and each transformation process of agricultural products will have intrinsic characteristics which will make them, always depending on spatiotemporal dimension (the where and when), more or less water absorbent.

Wine is obtained from a product of agricultural origin—the grape, through a transformation process. The virtual water content of processed products originating from agricultural products, therefore, includes the virtual water content of these, as well as the quantity of water required to complete the production process. Therefore, to calculate the virtual water for a bottle of wine, it is necessary to begin with an analysis of the volumes used in the vineyard to produce the grapes, to which must be added the volumes used in the cellar to transform the grapes into wine. To this "direct" consumption, that is, the consumption the producer company is directly responsible for, in that it is a controlled consumption occurring on company premises, we need to add the indirect consumption, that is the water consumption for the production of the inputs required to produce the end product. In the case of wine, then, it would be the agri-pharmaceuticals and fertilisers used in the vineyards, enological additives and packing products used in the cellar, and the energy consumption and fuels used for the entire life cycle of the wine bottle (from the vineyard to the selling of the product). In calculating the water footprint, the water consumption for each step identified in the wine bottle's life cycle is broken down based on the type of water used to give the water footprint for each step, and from the sum of this the total, water footprint is obtained (Hoekstra et al. 2011).

3 The Virtual Water in a Glass of Italian Wine

Let us now discover the hidden water in a glass of Italian wine. Even if by adding "Italian", it can appear rather unnecessary, in reality, it is not. In fact, the water footprint is spatiotemporal explicit—the volumes of water used change depending on *where* and *how* the wine is produced. However, not everyone knows that Italian wine has a lower water footprint compared to the average global value. Very often, the information concerning this is not as clear as it should be and the water footprint appears as a number not only far from the understanding of people but also from reality. Let us try to clarify the situation a little. From the database that the Water Footprint Network began constructing in 1996, we can find the data on the water footprint values of a wide range of agricultural products and their derivatives, of biofuels, and of products originating from livestock breeding or industrial origin (Mekonnen 2010; http://www.waterfootprint.org/?page=files/WaterStat-Product WaterFootprints). Each of these is accompanied by, in different detail, its geographic position (nation or region) and the indication of the volumes of water used subdivided by water typology. The water footprint of wine, globally, nationally and regionally, is shown in Table 1. If we compare the data, we can see how the Italian national average is more than 20 % lower than the global average.

A glass of wine requires 108 l of water broken down into 76 l of green (about 70 %), 17 of blue (about 15 %) and 15 of grey (about 15 %). If we look at the average national figures, we see how it is lower at 88 l, 67 (78 %) being green water, only 6 l (7 %) blue water and 15 l (15 %) grey water. The differences among the Italian regions reflect the different climate conditions, but also the different production strategies based on a lower or higher irrigation use, while the grey water value remains constant, practically equal to the world average. Looking at the data led us to considering the idea of studying more closely the question of the water footprint in the viniculture sector as an indicator for environmental performance. What is the significance of the green water footprint? Can the value of grey water be so unvarying in different national and international situations? Why are the blue water volumes in Italy on an average lower than the global one? An aspect that further interested us was to see how much the average data reported for statistical purposes in the database responded to the company situation, or if, instead, each company depending on its geographic and management features, distinguished itself significantly from the others.

4 Discovering Green Water

Water is an essential element for the life of plants. It is necessary for the carrying out of numerous chemical processes which occur in the plant tissue (e.g. chlorophyll photosynthesis), it allows for maintaining the cell turgor determining the

Table 1 The water footprint of wine

Liter/Liter of wine	World Average	National Average	Italian							
			Abruzzo	Basilicata	Calabria	Campania	Emilia-Romagna	Friuli-Venezia Giulia	Lazio	Lombardia
Green (l/l of wine)	607	534	536	584	583	584	502	519	568	503
Blue (l/l of wine)	138	46	33	67	74	41	57	3	43	33
Grey (l/l of wine)	124	117	120	133	139	125	108	103	120	103
WF (l/l of wine)	869	697	689	784	796	750	667	625	731	639

Liter/liter of wine	Italian									
	Marche	Molise	Piemonte	Puglia	Sardegna	Toscana	Trentino-Alto Adige	Umbria	Valle d'Aosta	Veneto
Green (l/l of wine)	569	569	505	549	565	565	430	567	406	499
Blue (l/l of wine)	20	39	17	126	28	28	0	38	0	24
Grey (l/l of wine)	121	122	104	128	123	123	103	121	103	104
WF (l/l of wine)	710	730	626	803	716	716	533	726	509	627

Liter/glass of wine	World Average	National Average	Italian							
			Abruzzo	Basilicata	Calabria	Campania	Emilia-Romagna	Friuli-Venezia Giulia	Lazio	Lombardia
Green (l/0,125 of wine)	76	67	67	73	73	73	63	65	71	63
Blue (l/0,125 l of wine)	17	6	4	8	9	5	7	0	5	4
Grey (l/0,125 l of wine)	15	15	15	17	17	16	14	13	15	13
WF (l/0,125 l of wine)	108	88	86	98	99	94	84	78	91	80
WF (l/0,75 l of wine)	648	528	516	588	594	564	504	468	546	480

Liter/glass of wine	Italian									
	Marche	Molise	Piemonte	Puglia	Sardegna	Toscana	Trentino-Alto Adige	Umbria	Valle d'Aosta	Veneto
Grey ((l/0,125 l of wine)	71	71	63	69	69	71	54	71	51	62
Blue (l/0,125 l of wine)	3	5	2	16	14	4	0	5	0	3

(continued)

Table 1 (continued)

Liter/glass of wine	Italian									
	Marche	Molise	Piemonte	Puglia	Sardegna	Toscana	Trentino-Alto Adige	Umbria	Valle d'Aosta	Veneto
Grey (l/0,125 l of wine)	15	15	13	16	17	15	13	15	13	13
WF (l/0,125 l of wine)	89	91	78	101	100	90	67	91	64	78
WF (l/0,75 l of wine)	534	546	468	606	600	540	402	546	384	468

WF glass (0.125 L)	WFN	A	B	C	D
	Italy	Centre	South	North	South
Green (l water/glass of wine)	67	60	55	69	159
Blue (l water/glass of wine)	6	6	1	1	1
Grey (l water/glass of wine)	15	63	0	1	23
Total (l water/glass of wine)	88	129	56	71	183

Source Mekonnen (2010)

rigidity and characteristic aspect of various parts of the plant, it guarantees the thermoregulation, and it allows for the transporting of nutritive substances within the plant. In fact, in the latter, the mineral elements present in the soil necessary for a normal development and growth of the plant species are dissolved. Water becomes a particularly important resource for the cultivation of arboreal and herbaceous species which make up the plant cycle in the spring–summer period, when rainfall is usually scarce. This fundamental resource is supplied to the crops, in part from the environment in the form of rain and in part by man using irrigation methods. Green water only involves the environmental part of the water required for the vines during the crop cycle (Hoekstra 2011).

The green water calculation is based on weather data, on soil and farming methods. This information is necessary to obtain the potential evapotranspiration value and from this calculate the effective evapotranspiration (from among the different methods to calculate the potential evapotranspiration, we used the Penman–Montheith modified by FAO; FAO 1998).

The assessment of the green water footprint leads to defining the quantity of rainwater held in the soil and used in the evapotranspiration processes during the crop cycle. For example, very often, rain falls during the winter, a time when the vine is dormant, while the vine itself requires water during the summer, a time when rain is generally scarce. In other words, the millimetres of water provided by winter rains do not all fall into the water footprint calculation. Indeed, only a very small part of this water quantity is effectively used by the vines as can be noted by studying the crop coefficient values for the vine during the different periods of the year (Table 2).

Table 2 Crop coefficient values of the grapevine in Italy

Month	Kc
January	0.2
February	0.2
March	0.2
April	0.48
May	0.59
June	0.68
July	0.68
August	0.68
September	0.68
October	0.59
November	0.38
December	0.2

Source FAO (1998)

Box. Water and the Vine: From Evapotranspiration to Water Stress

Evapotranspiration is the quantity of water (expressed by unit time) which from the soil-plant system passes to the vapour state due to the combined effect of transpiration, through the plant, and evaporation from the ground. Two distinct and different phenomena are involved which, however, are neither logical nor possible to separate. The evaporative part is, in fact, not closely linked to the crop, but however affects the water balance of the soil-plant system and from a practical point of view goes to determining the quantity of water effectively consumed.

The potential evapotranspiration is the maximum quantity of water which can be lost during the unit time for the evaporation and transpiration by 1 m^2 of a *Festuca arundinacea* field in an optimised and standardised condition. Therefore, the value is independent of the crop and the methods carried out, but, instead, depends on the seasons and the climate influencing the atmosphere's evaporating power. For this reason, for each crop there exists empirically determined "crop coefficients", differentiated by the phenological phases and subject to changes due to environmental factors (farming method, climatic region, etc.). The phenological phases identify the different stages of the life cycle of a plant species characterised by a morphological status to which specific physiological needs are linked. When possible, crop coefficients related to their specific climatic and geographic region should be adopted and, in the absence of the relevant documentation, reference should be made to similar environmental conditions. The crop coefficient value may be less or more than the unit and must be multiplied by the potential evapotranspiration to provide the effective evapotranspiration. Therefore, depending on the crop coefficient value, the effective evapotranspiration may be less or more than the potential evapotranspiration. In non-standard climatic and management conditions, the estimate of the effective evapotranspiration can be improved ("adjusted") by introducing a stress coefficient, *K*s. This is a water stress coefficient. In fact, while the effective evapotranspiration represents the crop's consumption when it is maintained with an optimum water supply where it can therefore transpire without limitations, in reality, the crops, even if irrigated, can encounter stress conditions which reduce their transpiration rate. Stress conditions can be verified when the soil's water content falls below the threshold value defined as "the easily usable reserve". Therefore, the crop is not in a condition of stress when the soil has a water content more than the easily usable reserve value. In these conditions, the stress coefficient will be 1 and the effective evapotranspiration value is not adjusted. Instead, the stress coefficient becomes less than 1, tending towards 0, when the soil water content falls below the easily usable reserve value. Therefore, the stress coefficient is maintained at 1 until reaching the value equal to the easily usable reserve, to then fall to 0, when the soil moisture reaches the wilting point value. The easily usable reserve value is typical of every type of soil and varies from crop to crop depending on the soil water withdrawal efficiency and the root system depth (FAO 1998).

There are those who criticise the fact that, in the virtual water volume estimates, this quantity of water naturally supplied by the environment by means of rain is also taken into account (Daniels 2011; Berger 2012). It is true that rainwater which is used by the vine would fall on the soil anyway, and it is true that any plant species would use it, but it is especially true that the vine always has a crop coefficient of less than 1, that is, it requires less water than the standard *Festuca* field, but also less water than a mixed oak forest where for 220 days (beginning from the 121st day, the 1st May), the crop coefficient value is more than 1 reaching a peak of 1.2 (INEA 2005).

The new and important point that the green water footprint raises is how much does a crop adapt or not to being cultivated in a given climate zone. When the green water footprint is high and significantly higher than the blue water footprint, then it is in a zone suitable for the production of grapes. Looking at this from an international trade perspective, the exportation of wine with a high green footprint to countries where the production of wine grapes requires a higher water volume for irrigation purposes and where, therefore, the possible or actual production of wine would be characterised by a higher blue water footprint value would be an advantage in terms of overall water resource saving!

Box. Green Water and False Myths: "The Green Water Footprint is Higher in the South!"

Often, those who approach the concept of the water footprint believe that it penalises the crops of warmer regions, where the potential evapotranspiration is higher. It is true—where the average temperatures are higher, potential evapotranspiration is higher, but this does not suffice. The green footprint does not tell us how much water the plant would require during the crop cycle to compensate for the potential evapotranspiration, but it tells us how much rainwater was held by the soil and used by the crop. Therefore, if, let us say, it rained during the year, much more than the vine required, the green water footprint would be equal to the effective evapotranspiration value and not to the quantity of rainwater which fell during the crop season. If, instead, as usually happens, the rainfall during the year was less than the water volume required to compensate for the potential evapotranspiration calculated for the crop cycle, the green water footprint would correspond to the quantity effectively used during the crop cycle provided by the difference between the effective daily rainfall and the quantity of water in excess that is not held in the soil and which percolates down (calculated in the absence of irrigation). The sum of these values during the crop period corresponds to the sum of the effective evapotranspiration values calculated under water stress conditions, and these values correspond to the green water.

Consequently, contrary to how it may seem, the green water footprint is not higher than in the south. In fact, in hotter and less wet regions, the green water footprint should be lower, whereas the blue water footprint is higher as irrigation methods are required to avoid, even for the highly resistant vine, qualitatively and quantitatively excessive and harmful water stress.

5 Discovering Blue Water

In the production of wine, the consumption of blue water occurs in different phases, beginning with the vineyard where the blue water is used for irrigation, for the dissolution and use of active principles used for protecting, and washing the agricultural machinery used, right up to the cellar where water is required in almost every step in the wine-making process.

As shown in Table 1, the blue water volumes are generally low in wine. This is due to two main reasons. As far as the vineyard is concerned, it should be remembered that the vine is a plant that has an extremely high resistance to water stress, a reason why it is often not irrigated and, even when irrigated, it tends to be an emergency irrigation using very low volumes of water compared to those required to compensate for the evapotranspiration (even less than 50 %; Pou 2011). Of the approximate 2.5 million irrigated hectares in Italy, only 7.4 % involves vineyards, about 30 % of the Italian vineyards (INES 2005). Instead, in the cellar, consumption is generally low and it is estimated around 5–8 l of water are used for each bottle of wine, highlighting how wine making is a highly technological and efficient process concerning water consumption (Winetech 2005; Di Stefano 2008).

Only the water used for irrigation methods is considered from the data, and it is in all likelihood a conservative value. In fact, where the grapevine is not irrigated, the value is significantly lower, and only includes the volumes relevant to agri-pharmaceuticals and the washing of the agricultural machinery (Table 1).

Box. Vine irrigation: Strategies and Methods

The grapevine is a traditionally non-irrigated crop able to tolerate summer season water stress, and thus, it is usually found in dry areas. In the past, the *Vitis vinifera sativa* was cultivated in low-growing rows of vines, and because of its extensive and deep root system, it never required irrigation. Modern viticulture, instead, must take into account the introduction of new more resilient hybrid grapevine varieties, growing taller and with more foliage spread, focusing more on the must quality, and finally, but not less important, the effects of the present climate change. Weather conditions can, in certain phenological phases, be insufficient to meet plant growth requirements, and if once grapevine irrigation was considered a "forced" agronomic practice that is aimed at maximising yield to the detriment of the end product quality, it is now permitted as an "emergency" practice under legislation, also of high-quality wines. Moreover, it should not be forgotten that in vine cultivation marked for wine production, an excessive artificial water input or one carried out after the maturation phase may be harmful in terms of must quality. For these reasons, the European Community has given the right to EU member states to establish their own regulations on the use or non-use of emergency irrigation in DOC and DOCG production, while specifically prohibiting along with some other methods the use of irrigation as a forced cultivation.

Among those irrigation methods widely used and able to respond to the needs of an "emergency" situation, "regulated deficit irrigation (RDI)" stands out, where irrigation must only be implemented in the most sensitive phases of the crop growth, or the flowering-setting phase and the period between the veraison and the ripening phase. A reduced or zero water input in other phases allows for reducing the plant canopy growth and increasing the "fruit to leaf" ratio, improving berry colour and quality resulting in the right skin/flesh ratio and an increase in the content of the anthocyan and phenolic elements (McCarthy 2002; Pou 2011). The irrigation methods that adapt better to this strategy are the localised ones. This is drip irrigation with drip water supply lines (above ground) or sub-irrigation (underground). These methods, which are more commonly used in viticulture, result in an efficient water resource use further justifying the low blue water footprint values (McCarthy 2002; Pou 2011).

6 Grey Water

The grey water footprint generally expresses a volume of water that has not been actually used, but the volume that would be required, if water body pollution occurred, to return the concentrations to below the legal standards or the established ecotoxicological end points. A new term coined in 2008 by Hoekstra and Chapagain, the grey water footprint, expressed a new and simple way to report pollution, generally expressed as a concentration (quantity in weight compared to volume) in units of volume and able, therefore, to be added to the green and blue footprints to provide a comprehensive water footprint (Hoekstra 2008). Thus, the grey water footprint is obtained from the pollutant load divided by the difference between the maximum acceptable concentration in the receiving water body and the natural background concentration. This because in the case where pollution had already been present in the water body, the water body's capacity to assimilate it would be consequently reduced. The maximum acceptable concentration depends not only on the pollutant but also on the water body it was found in. There are different legal and ecotoxicological standards according to the type of water and that water's use. Surface water and groundwater apart from the differences in the maximum acceptable concentrations are, instead, considered as a single water body when calculating the dilution volumes concerned. In calculating these, furthermore, the critical pollutant is used, that is the dilution volumes are not summed up but, instead, the highest is selected (that would be required for the pollutant based not only on the concentration reached in the water body but also on its toxicity as per the legal and ecotoxicological standards) as that virtual volume would also dilute the other pollutants, which would require lower dilution volumes (Hoekstra et al. 2011).

In order to calculate the grey water volumes, the manual itself proposes three types of approaches, from the simplest, to the more complex and realistic.

1. The simplest approach uses a fixed fraction to express the part of the pollutant which reaches the water body. This index depends on the type of pollutant and comes from the sector literature.
2. The second approach uses standardised and simple models to forecast the pollutant concentrations found in the water body.
3. The last approach, instead, is proposed as a refining of the second where sophisticated and complex models are used able to forecast the pollutant concentrations in the water body based on the company scenario (Hoekstra et al. 2011).

Table 1 shows the similar grey water footprint values in Italian regions compared to the world average. Why?

In order to estimate the grey water footprint, the database was actually created using a Type 1 approach which only includes nitrogenous fertilisers and estimates a nitrate leaching of 10 %.[1] The maximum acceptable concentration used is the legal standard for nitrates (50 mg/l) and assumes a natural groundwater concentration of 0. The nutritive input for a vine is considered practically the same worldwide and in the Italian situation, and thus, the grey water footprint remains constant. Moreover, it has been evaluated that the nitrogenous fertilisers are the critical pollutants, and therefore, no agri-pharmaceutical application is taken into consideration.

This mean data obviously involve many different situations. Fertilisation is not always adopted as a practice just as sometimes it is not a given that the nitrates are the critical pollutants. The generalisation can be made, but it does not always correspond to the mean value, and sometimes does not even represent the "worst case scenario". Let us look at why.

---

[1]The term leaching refers to the transporting of dissolved solutes in water, such as nitrates and agri-pharmaceuticals, which penetrating beyond the root system depth reach and pollute the groundwater. The nitrates, being highly soluble and mobile in the soil, are a risk of water pollution in that the excess not employed by the plants is washed away by rainwater or irrigation water and can reach, through leaching, the deeper water bodies.

We attempted to calculate the grey water footprint of some Italian wine-producing companies different in their management strategies and their geographic position. However, firstly, we made some of our own modifications to the model:

- we decided to use as a fixed nitrate leaching percentage, not 10 % but 6 %. This is a value obtained after 15 years of studies using a lysimeter in Europe that seemed for us to respond better to the actual situation (Fank 2006);[2]
- we also considered the use of agri-pharmaceuticals, and by means of scientifically validated models, we assessed the percentage which reached the surface and groundwater bodies, applying, in the first case as the maximum acceptable value, the No Observed Effect Observation(NOEC) and, in the second, the legal standard for groundwater as established by the European Water Framework Directive (EU 2000). In addition, we also considered the possible mitigating measures adopted by the companies resulting in reducing their pollution levels.

Consequently, for each agri-pharmaceutical, we obtained a dilution volume for the groundwater and for the surface water and we considered the higher of the two. Therefore, from among all the agri-pharmaceuticals applied in the same vineyard, we took as the grey water footprint of the agri-pharmaceuticals applied, the value relative to the resulting higher dilution volume. We then compared this value with the fertilisation value, and again, we chose the highest, thus, obtaining the grey water footprint, that is the water volume that would have diluted the pollution resulting from the use of nitrogenous fertilisers and agri-pharmaceuticals.

The results change depending on the active principles used, but also on the company's location in relation to surface water bodies. For companies with vineyards more than 100 m from a surface water body, the pollution of this body becomes highly unlikely, while this becomes very significant for those vineyards found far from a water body.

The values are sometimes above and sometimes below what is reported in the database. In company A, the main differences are found in the grey footprint. This company uses irrigation and is positioned near a water body, and the latter information explains the high grey water footprint. It is considered, in fact, that a moderate amount of pollutants may reach the water body requiring a high dilution volume (more than 60 l/glass for the critical pollutant, this involves a volume that also dilutes the other pollutants) (Fig. 3).

[2]The lysimeter is a device used to assess groundwater dynamics. It is equipped with a drainage system to collect the percolated water. The study mentioned analysed the percolated water for a period of 15 years. On the basis of the findings, it seemed opportune to reduce the fixed nitrate leaching percentage.

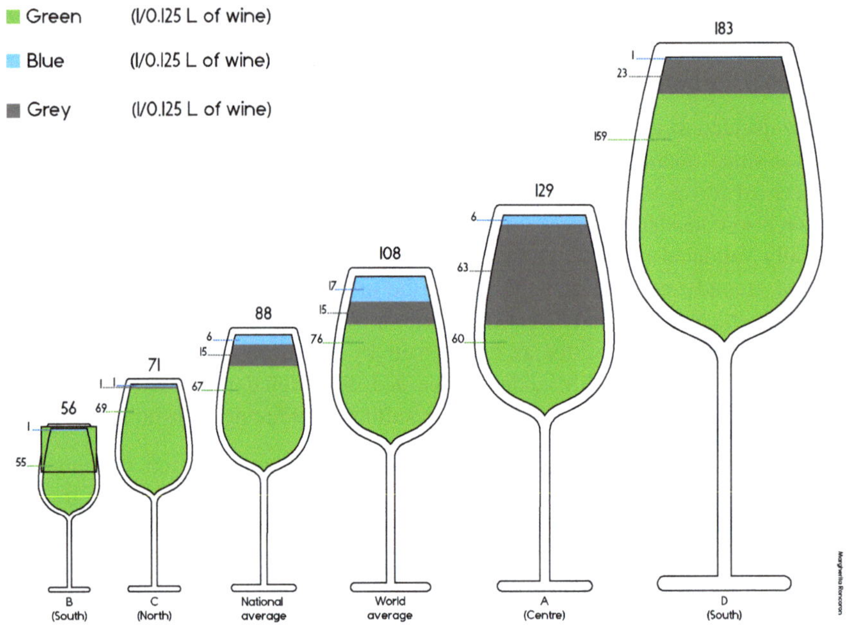

Fig. 3 The water footprint of Italian wine in 4 different situations. The results are compared with the mean data and expressed per glass of wine (0.125 l)

WF glasses (0.125 l)	NWF* Italy	A Centre	B South	C North	D South
Green (l water/glass)					
Blue (l water/glass)					
Grey (l water/glass)					
Total (l water/glass)					

* Data from water footprints of crops and derived crop products (1996-2005). Report 47, Appendix II.

In company B, irrigation methods are not used, no nearby water bodies exist, and no use of fertilisers. These factors result in markedly low blue and grey water footprint values.

In company C located in the north, irrigation and fertilisers are not used and the nearest surface water body is 200 m away. Despite this, a significant part of an active principle used in high amounts (copper oxychloride, 3.2 kg/ha) reaches the water body, thus, necessitating dilution.

The last company, D, is located in the south of Italy and shows the highest green value which, however, is due to the low production yields (50 cwt/ha). The blue value, however, remains low, while the grey water footprint results from the virtual dilution after fertiliser application.

Therefore, it is important to note how low yields penalise the products, demonstrating a low efficiency in water resource use (green, blue and grey). If, in fact, from the same planted surface area, higher grape yields were obtained and, therefore, a higher quantity of wine, the water footprint values would be lower. Therefore, using less water in production, from a water efficiency point of view, is an assumption of sustainability, even if, for qualitative reasons or due to particularly adverse environmental conditions, and it often appears to be a choice that is not always very practicable.

Box. Sustainability and Quality: Synonyms or Antonyms?
It is often believed that high-quality wines are more environmentally sustainable than those of lower quality. But the environmental indicators, including the water footprint, sometimes tell us the opposite. Why? Maybe we should take a few steps back and clarify our ideas. First of all, the water footprint of a lower quality wine could in some cases be lower to that of a higher quality wine due to differences in yields per hectare. Those who must produce a cheaper wine will try to maximise the yields in order to contain costs, on the contrary, those aiming for quality will have to select better grape clusters during the crop season to be able to reach the established qualitative standards. It is clear that those, with the same weather conditions and cultivation techniques, from the same farmed surface area and a higher grape yield and, therefore, more wine will have an overall lower water footprint. In fact, the green, blue and grey water volumes used will be distributed over a higher product quantity.

However, we should stop here to reflect if this comparison is logical and correct. The environmental impact indicators from a sustainability perspective should not lead to comparing the end products, but should be used to objectively measure product improvements from the same situation following given choices. Therefore, in the case of wine, the use of indicators should be used by the winery interested in reducing the impact on water resources to assess the initial state, to identify the critical areas, to plan effective strategies to reduce the impact in delineated critical areas and, finally, to measure the improvements achieved. Consequently, the intra-company and not the inter-company comparison becomes important. The fact itself that even hypothesising a similar management strategy in the field and in the cellar in different company situations due to geographical position, climate conditions and territorial context have quite different water footprint values, should discourage us from any attempt to compare, which, however, remains interesting for statistical purposes in creating global, national and regional databases. Lastly, but not of less importance, the fact would remain that the two wines hypothesised (quality wine obtained from low yields due to management choices in selecting better grape clusters and lower quality wine obtained from vineyards aimed at maximising yields) respond to two different needs and cannot be compared, not only for their environmental impact, but also for their intrinsic qualitative characteristics, for their economic value and their market positioning. They respond to different needs, pursue different objectives, require different management strategies and, therefore, result in products with different targets and so any comparison loses every possible significance.

Box. Communicating Virtual Water It is definitely not easy. It often involves high numbers, sometimes incredibly high. Who would not be surprised to know that for a good glass of wine, it requires, on an average, 109 l of water when we actually pour only 0.125 from the bottle (872 times less!)? We can easily imagine that there is "hidden" water, but it would be much more difficult to calculate the right weight.

For this reason, a simple number should not be delegated to carry out this task. How many times have we read 3000 l for a 200 g beef steak, but how many times have they told us that 94 % of the volume of water is green water, 4 % is blue and 3 % grey? The values should always be shown by water typology, and the different water typologies should always be explained. It is quite different knowing that of the 109 l of water required at a global level for producing one glass of wine, only 17 are of blue water, 15 grey and 76 of green water. Moreover, as we have seen, on a world average, Italian wine has an even lower water footprint, a reason why the information on the product should not only be clear, referring to the single elements of the water footprint, but also linked to the spatiotemporal explication. Thus, the water footprint would include information and not a number, far from our understanding and also from the existing reality.

7 Some Drops of the Wine Water Footprint

The virtual water and water footprint of wine express the "hidden" volumes of water, calculated from the vineyard to the cellar, required in producing the wine. The water footprint, extending the virtual water concept, distinguishes between the volumes depending on the water typology used—green water is the rainwater effectively used by the crops, blue water is the surface water and the groundwater withdrawn and used in the production processes, and grey water is the volume of water required to neutralise the resulting pollutants. In Italy, the recognised value is 88 l per glass of wine (0.125 l), but the findings of experiments conducted in different national contexts result in values varying between 56 and 183 l depending on the management and geographic aspects of the wineries involved.

Almost all the "hidden" water in a glass of Italian wine is green water (55–159 l per glass of wine corresponding to a percentage varying from 46 to 97 %). It is important to note here that a cultivation of *Festuca arundinacea* (a grass commonly used in lawns) or a wood of oak trees would require much more green water than the grapevine itself. As far as blue water in Italy is concerned, it makes up a very small percentage of the total, varying between 0.5 and 7 %. It should be recalled that the *Vitis vinifera sativa* is able to tolerate high water stress levels and that Italian cellars are equipped with a high technological efficiency in limiting the water consumption required in the processing stages.

From the general data, a certain consistency in grey water volumes appears (approx. 15 l per glass) and by trying to assess, through scientific models, the pollution levels of water bodies based on the management strategies used, and of the company scenario (also including the company carrying out mitigation measures), different values are obtained, varying from 0 to 63 l per glass of wine! These data accurately mirror the studied product's water footprint and can help companies in reducing their impact through modifying their management strategies.

For these reasons, the water footprint of a glass of wine should always be presented based on the water typology used and on the spatiotemporal position of the wine company. Indeed, the hidden volumes can often seem "enormously" high, especially to those not actually involved in the production. Being clear and transparent would help in conferring the right weight to water, especially the water we do not see!

Bibliography

Berger M., & Finkbeiner, M. (2012). Methodological challenges in volumetric and impact-oreinted water footprints. *Journal of Industrial Ecology*. doi: 10.1111/j.1530-9290.2012.00495.x.

Coca-Cola Europe. (2011). *Towards sustainable sugar sourcing in Europe. Water footprint sustainability assessment WFSA*; http://www.waterfootprint.org/Reports/CocaCola-2011-WaterFootprintSustainabilityAssessment.pdf.

Daniels, Plenzen M., & Kenway, S. J. (2011). The ins and outs of water use—A review of multi-region input–output analysis and water footprints for regional sustainability analysis and policy. *Economic Systems Research, 23*(4), 353–370.

Di Stefano, N., et al. (2008). *A low cost land based winery wastewater treatment system: Development and preliminary results*. Report v: CSIRO Land and Water Science. 43.

EU. (2000). *European water framework directive* n. 2000/60/CE; http://www.directivaacque.minambiente.it/.

Falkenmark, M., & Widstrand, C. (1992). Population and water resources: A delicate balance. *Population Bulletin, 47*(3), 1–37.

Fank J. (2006). Die Bewirtschaftung des Versuchsfeldes Wagna—Auswirkung auf die Grundwassersituation", Seminar Umweltprogramme für die Landwirtschaft und deren Auswirktung auf die Grundwasserqualität, HBLFA Raumberg-Gumpenstein, Research Report European Lysimeter platform.

FAO. (1989). *Crop evapotranspiration. Guidelines for computing crop water requirements*. FAO irrigation and drainage n. 56; http://www.fao.org/docrep/X0490E/X0490E00.htm.

FAO. (1998). *Crop evapotranspiration. Guidelines for computing crop water requirements*. FAO Irrigation and Drainage no. 56; http://www.fao.org/docrep/X0490E/X049oE00.htm.

FAO. (2012). *Coping with water scarcity. An action framework for agricultural and food security*, Water reports 38; http://www.fao.org/docrep/016/i3015e/i3015e.pdf.

Galli, A., et al. (2012). Integrating ecological, carbon and water footprint into a 'footprint family' of indicators: Definition and role in tracking human pressure on the planet". *Ecological Indicator, 16*, 100–112.

Hoekstra A. Y., & Chapagain A. K. (2008). *Globalization of water: Sharing the planet's freshwater resources*. New York: Wiley.

Hoekstra, A. Y., et al. (2011). *The water footprint assessment manual: Setting the global standard*, Earthscan; http://www.waterfootprint.org/?page=files/WaterFootprintAssessmentManual.

Inea—Istituto nazionale di economia agraria. (2005). *Uso irriguo dell'acqua e principali implicazioni di nature ambientale*; http://www.inea.it/pdf/USOIRRIGUO2.pdf.

IWMI. (2007). *Water for Food Water for Life*. USA: Earthscan.

McCarthy, M. G. et al. (2002). Regulated deficit irrigation and partial rootzone drying as irrigation management techniques for grapevines. In *Deficit irrigation practices* (no. 2, pp. 79–87). Water Reports FAO.

Mekonnen, M. M., & Hoekstra, A.Y. (2010). *The green, blue and grey water footprint of crops and derived crop products: Volume 1 Main Report*, Value of Water Research Report Series n. 47, UNESCO-IHE;http://www.waterfootprint.org/Reports/Report47-WaterFootprintCrops-Vol1.pdf.

Mekonnen, M. M., & Hoekstra, A. Y. (2011). *National water footprint accounts: the green, blue and grey water footprint in production and consumption: Volume 1 Main Report*. Value of Water Research Report Series no. 50, UNESCO-IHE;http://www.waterfootprint.org/Reports/Report50-NationalWaterFootprints-Vol1.pdf.

Molden, D., & de Fraiture, C. (2010). Comprehensive Assessment of water management in agriculture. *Agricultural Water Management, 97*(4), 493–578.

Pou, A., et al. (2011). Cover cropping in Vitis vinifera L. cv. Manto negro vineyards under Mediterranean conditions: Effects on plant vigour, yield and grape quality. *Journal International des Sciences de la Vigne et du Vin, 45*(4), 223–234.

Prüss-Üstün, A., Bos, R., Gore, F., & Bartman, J. (2008). *Safe water, better health*. USA: WHO.

Revenga, C. et al. (2000). *Pilot analysis of global ecosystems: Freshwater systems*. World resources institute; http://pdf.usaid.gov/pdf_docs/Pnacs566.pdf.

Rijsberman, F. R. (2006). Water scarcity: Fact or fiction? *Agricultural Water Management, 80* (1–3), 5–22.

Rosegrant, M. W. et al. (2002). *World water and food to 2025: Dealing with scarcity*, International Food Policy Research Institute; http://www.ifpri.org/sites/default/files/pubs/pubs/books/water2025/water2025.pdf.

Shiklomano, I. A. (1993). World fresh resources. In: P. H. Gleick (Ed.), *Water in crisis: A guide to the world's freshwater resources*. Oxford: Oxford University Press.

Shiklomanov, I. A. (1999). World water resources: Modern assessment and outlook for the 21st century. *GEO-3—Global Environment Outlook* 3, UNEP.

UNDP. (2006). *Human Development Report 2006. Beyond scarcity: Power, poverty and the global water crisis*; http://hdr.undp.org/en/media/HDR06-complete.pdf.

UNESCO. (2009). WWDR3: *Water in a changing world*, The United Nations World Water Development Report 3. USA: Earthscan.

University of California Division of Agriculture and Natural Resources. (1989). *Irrigation scheduling: A guide for efficient on-farm water management*, Publication 21454.

Winetech. (2005). *Guidelines for the management of wastewater and solid water at existing wineries*.

The Water Footprint and Environmental Sustainability of Italian DOP, DOC and DOCG Food Products

Maria Cristina Rulli, Arianna Veroni and Renzo Rosso

The Italian "quality label" food product is synonymous with high quality and is very important from both an economic and cultural point of view.

The recent debate on the environmental sustainability of food products was the motivation behind this study where, using the water footprint indicator, some high-quality Italian food products were analysed concerning their impact on the environment, and particularly on water resources. Specifically, the water footprint was calculated considering its three green, blue and grey components, analysing the entire production stages where, in the case in which it was directly or indirectly of agricultural origin, the water footprint was determined by means of an agro-hydrological model in order to study the crop growth cycle and calculate the evapotranspirated water. These analyses considered the entire crop growth cycle and the product's processing phase in compliance with the product specifications and rules, where the water footprint is evaluated using the model parameterisation based on product's area of origin. Special attention was also placed on evaluating the impact of agronomic growth forcing—primary (irrigation) and secondary (fertilisation)—in calculating the water footprint.

The products analysed were the cheese, Parmigiano Reggiano DOP; three wines, Barolo DOCG, Sassicaia DOC and Moscato di Pantelleria DOC; and three types of Saffron DOP, Saffron from Aquila, Sardinia and San Gimignano.

The analysis of the results and, specifically, the model scenarios where it can be conjectured that the primary good was cultivated outside its zone of origin or with methods that deviated from the prescribed standards show how the quality label, linking the product to a well-delineated geographic area of origin, not only assures a high quality, but also environmental sustainability. In fact, a quality label product's zone of origin is marked by optimal climatic, topographical, soil and hydrological conditions guaranteeing the quality and specific features of products which comply with the restrictive limitations on growth forcing (irrigation and fertilisation).

M.C. Rulli (✉) · A. Veroni · R. Rosso
Department of Civil and Environmental Engineering, Politecnico Di Milano, Milano, Italy
e-mail: mariacristina.rulli@polimi.it

© Springer International Publishing Switzerland 2015

M. Antonelli and F. Greco (eds.), *The Water We Eat*,
Springer Water, DOI 10.1007/978-3-319-16393-2_17

1 Introduction

Recent economic development has contributed to the current marked imbalance in the Earth's natural systems where the scarcity of freshwater together with the growing demand for water has been highlighted. The common concept of the term "water users" has always been accepted as "those who use water for a purpose". Therefore, traditionally, water resource management policies have been addressed to these users. Recently, however, it was shown (Hoekstra and Chapagain 2007, 2008) that any water resource management approach that is based on these water users is quite limited. In fact, any such approach ignores the end consumers, the suppliers and all those companies involved in the supply chain. All uses of water, at the end, are tied to consumer end consumption. Consequently, it would be very interesting to know the water requirements and specific impact of consumer goods. This is particularly so for goods that do not specifically require a water input in their end-use phase (food products, bioenergy, natural fibre, etc. Aldaya and Hoekstra 2010).

Over the last decades, an environmental awareness has been spreading in the hope of creating a synergy in economic, social and environmental development. Through analysing a product's life cycle, environmentally sustainable strategies can be developed. This research work has been developed along these lines, using as examples well-known and widely consumed agricultural products, the objective being to show the impact of their production processes on water resources. Accurate information on the water footprint of goods can be an aid in drawing up strategies aimed at a more sustainable and equitable use of water resources.

To assess the impact of each production step on the available water resources, an indicator devised by Chapagain and Hoekstra (2003) is used—the water footprint. It is an indicator of water use which assesses both the direct and indirect consumption of a good's user or producer. The water footprint can be calculated for groups of well-defined consumers (e.g. an individual, a family, a town, a city, a province, a state or a nation) as well as for producers (e.g. a public entity, a private company or an entire economic sector) and for single products (Aldaya and Hoekstra 2010).

A product's water footprint is the volume of freshwater used to produce that good, measured at the site where that good was actually produced (Hoekstra and Chapagain 2008).

It refers to the total water volume used in the different production stages and is calculated in terms of the water volume used and/or polluted per unit of time. This indicator takes into account the green water footprint (rainwater infiltrated into the soil and taken up by plants), the blue water footprint (groundwater and surface water withdrew for the production of goods and services) and the grey water footprint (volume of water polluted through the production of goods and services). The latter is quantified as the quantity of freshwater required to assimilate the pollutant load based on the established quality water standards. In areas of water scarcity, the knowledge of the water footprint of a good or service can be useful in optimising the use of the scarce resources available (Aldaya and Hoekstra 2010).

It is important to establish whether the water used for the production of a good derives from the use of rainwater (green water) or the withdrawal of surface or ground water (blue water). The subdivision of these two indices is useful as the impact of a single product on the water resource can be assessed.

The scientific literature shows the water footprint calculation of some products, referring to their specific place of origin, such as cotton, milk, coffee, tea, tomatoes, bioethanol, and biodiesel (Chapagain et al. 2006; Hoekstra and Chapagain 2007; Chapagain and Orr 2009; Gerbens-Leenes et al. 2009).

In this paper, a water footprint analysis of only the green and blue components is presented for some Italian food products of great interest culturally and recognised at both a national and international level. The common feature of all the products studied is the recognition of their denomination of origin and geographic indication. They are entitled to use the labels DOP, DOC or DOCG—"quality labels" which respectively stand for "Protected Denomination of Origin", "Controlled Denomination of Origin" and "Controlled and Guaranteed Denomination of Origin". The quality labels distinguish the product based on its production conditions and controls established by quality standards and not on the identity of the producer company. This work has specifically focused on the fact that this was a characteristic of the food products studied as it disciplines and regulates all the production steps by means of specific product quality standards.

The food products studied, with the different above-mentioned quality labels, belong to the food categories of cheese, wines and spices. For the cheese category, Parmigiano Reggiano DOP was chosen; for the wines Barolo DOCG, Bolgheri Sassicaia DOC and Moscato di Pantelleria DOC; and for the DOP spices, saffron from the L'Aquila, Sardinia and San Gimignano areas.

The choice in analysing these products resulted in the following:

(a) calculating the effective water footprint;
(b) analysing the water footprint according to scenarios using irrigation;
(c) analysing the water footprint according to scenarios using fertilisation.

This choice was made based on the need to analyse the existing relationship between a product's water footprint and its zone of production. The fundamental requisite of the DOP, DOC and DOCG products is to guarantee the production zone, and consequently, they are particularly useful for an analysis of this type.

2 Materials and Methods

Parmigiano Reggiano cheese figures as one of the DOP products analysed. The specification governing this product identifies its production zones as the provinces of Parma, Reggio Emilia, Modena, Bologna and Mantua. This cheese is produced from the milk of cows fed mainly on fodder produced in the zone of origin and whose diet is prescribed in the product specifications and rules (the so-called *disciplinare del Parmigiano Reggiano*). Where the wine is concerned, three different grapevines and winery zones were identified, being defined by their respective product specifications—the production zone of Barolo in the Langhe Hills in Piedmont, Sassicaia from the Bolgheri area in Tuscany and Moscato di Pantelleria from the island of Pantelleria in Sicily. These areas have three different climates where the grapes are cultivated to produce the quality wines of Barolo, Sassicaia and Moscato di Pantelleria, following the specific production standards for the wine grape varieties of Nebbiolo, Cabernet Sauvignon (>80 %) and Zibibbo, respectively.

For the spices, the DOP label saffron was studied, a traditional and niche production in Italy, for three saffron production zones—L'Aquila, Sardinia and San Gimignano.

The database used for the analyses was created based on a careful study of the production standards of the products chosen, on agronomy manuals (Bianchi et al. 2007), interviews with producers and on current literature. The climate information for the production zones of the studied products was taken from the ARPA Websites, referring to a sample of 20 years for a total of 13 weather stations, while the pedologic data was taken from regional pedologic maps.

The water footprint calculation was carried out using the CROPWAT 8.0 model (Allen et al. 1998) following the procedure to estimate the water footprint drawn up by Hoekstra and Chapagain (2008) and Hoekstra et al. 2009. CROPWAT 8.0, for its relevant crop, climate and pedologic features, estimates the net rainfall according to the SCS-CN method (USDA 1985) and determines the water requirement of the specific crop using the FAO Penman–Monteith method. By means of this method, firstly, the evapotranspiration for a reference crop is estimated; then, using the crop coefficient for the specifically studied crop, its effective evapotranspiration is determined. In particular, the reference conditions indicate a crop (height of 0.12 m; surface resistance of 70 s m^{-1}; albedo of 0.23) which grows in large areas, with excellent agronomic conditions and water availability meeting crop needs.

Once the crop's water requirement is calculated, it is compared with the water availability in the studied area, thus determining the water footprint of the agricultural product, eventually broken down into the green water footprint and blue water footprint elements. The green water footprint is the quota of the water requirement satisfied by the effective rainfall, and the blue water footprint is the quota of water that reaches the crop through irrigation. The water footprint calculation is shown in detail in the *Water Footprint Manual* (Hoekstra et al. 2009).

When the food product of which the water footprint is to be determined is not primary (as we will see in the following case of Parmigiano Reggiano), the water footprints of all the primary products that make up the end product will be estimated and then summed up proportionally.

3 Case Study

3.1 Parmigiano Reggiano Cheese

The water footprint calculation of Parmigiano Reggiano cheese uses as a reference the milk produced by a "Reggiana breed" cow whose diet is governed by the Parmigiano Reggiano specifications and rules that are collected in the so-called disciplinare del Parmigiano Reggiano. In the chapter named "Regulation on the feeding of bovines", "disciplinare del Parmigiano Reggiano" specifies that "in feeding dairy cows,

- at least 35 % of the dried feed used must be of company production;
- at least 75 % of the dried feed must originate from the Parmigiano Reggiano area".

In this study, the dairy cow feed was considered as that originating entirely from the production zone, specifically referring to the Parma area. In compliance with the established specifications, a dairy cow's diet was considered as being made up of 11.5 kg of hay, 7 kg of maize and 4.5 kg of barley per day. Moreover, it was assumed that a cow consumes 40 l of water a day. The water footprint of the single feed items making up the cow's diet, calculated using the meteo-climatic and pedologic data of the same farming area, affects the water footprint of Parmigiano Reggiano cheese. To estimate the water volume required by each crop, the CROPWAT 8.0 model (Allen et al. 1998) was used; then, based on the average crop yields in the production zones, the water footprint for each crop was estimated (Table 1).

As far as crop irrigation is concerned, the product specifications do not prescribe any specific method, consequently an irrigation on demand is assumed. From this, the blue water footprint values are calculated. The grey water footprint has not been calculated in this study.

The water footprint components of the single feed items were then compared with the daily feed ration, thus obtaining the footprint of a Reggiana breed dairy cow's diet for the production of Parmigiano Reggiano:

$$3100 + 2835 + 4800 = 10.735\ \text{l/day}\ \ (\text{green water footprint})$$

$$9 + 735 + 40(\text{water consumed}) = 784\ \text{l/day}\ \ (\text{blue water footprint})$$

A Reggiana breed dairy cow produces on an average 16 kg of milk per day, while a Parmigiano Reggiano cheese form requires an average of 550 l of milk, and after 24 months of maturing weight about 40 kg (the product specification for Parmigiano Reggiano). Finally, knowing that one litre of milk weighs about 1.030 kg, we can estimate the water footprint (Table 2).

Table 1 Estimate of the water footprint of a dairy cow's daily feed ration for the production of Parmigiano Reggiano

		Water requirement m^3/ha	Yield t/ha	Water footprint m^3/t	Dairy Cow's diet (kg/day)	Water footprint (l/day)
Barley	Water footprint $_{green}$	2.620	3.8	689	4.5	3.100
	Water footprint $_{blue}$	8		2		9
Maize	Water footprint $_{green}$	3.874,5	9.5	405	7	2.835
	Water footprint $_{blue}$	997,5		105		735
Fodder	Water footprint $_{green}$	4.327	10	433	11.5	4.800
	Water footprint $_{blue}$	–		–		–

Source edited by the authors

Table 2 Estimate of the water footprint for 100 g of Parmigiano Reggiano

	Water footprint of a 40 kg of Parmigiano Reggiano form (l)	Water footprint of 100 g of Parmigiano Reggiano (l)	
Water footprint $_{green}$	380.086	950	1.020
Water footprint $_{blue}$	27,761	70	

Source edited the authors

3.2 *Wines and Viticulture: Barolo, Sassicaia, Moscato di Pantelleria*

For each of the viticulture productions, four reference rain gauges were selected and according with the monographs to the selected grape variety (Soster and Cellino 1998), the starting of growing period for each grapevine variety analysed was identified (the first week of April for Barolo, the end of April for Sassicaia and the second 10 days of April for Moscato di Pantelleria).

The specifications and rules for the production DOC and DOCG wines (the so-called disciplinare) prohibit the irrigation. Therefore, the water footprint of each wine grape variety is basically a green water footprint which is obtained by comparing the sum of the water volume required to complete the plant growth cycle and the effective rainfall, all scaled to the crop yield. This water volume was calculated on decadal base using the CROPWAT 8.0 model (Allen et al. 1998; FAO 2009) and then summed up on the plant growth period. From the disciplinare of the specific wine considered, the maximum agricultural yield of grapes per hectare and the maximum yield of wine from grapes (in percentages) can be individuated. From the latter, the litres of wine produced can be calculated (Table 3).

As can be seen in Table 3, the highest water footprint is that of the Moscato di Pantelleria wine, while the Barolo and Sassicaia footprints are comparable. This is due to the different role played by the hydrological phenomena in the hydrological balance. On the island of Pantelleria, where the Moscato di Pantelleria from the Zibibbo grape is produced, the high evapotranspiration, experienced especially in

Table 3 Estimate of the water footprint for a litre of wine and a 0.75 l bottle of wine

	Crop water requirement (cwr) versus water availability (m^3/ha)	Max. yield l_{wine}/ha	Water footprint of 1 l of wine (l_{water}/l_{wine})	Water footprint of 1 bottle of wine ($l_{water}/0.75l_{wine}$)
Barolo	$2.726_{cwr}/100\ \%_{available}$	5.600	487	365
Sassicaia	$3.605_{cwr}/100\ \%_{available}$	7.200	500	375
Moscato di Pantelleria	$5.096_{cwr}/65\ \%_{available}$	6.000	548	410

Source edited by the authors

the summer months, is not compensated by an adequate rainfall. The plant's water requirement is greater than the green water volume available for the plant. The Zibibbo grape variety would require more water to reach its full growth and vigour, but the DOC and DOCG specifications and rules require the absence of irrigation. Despite this, the traditional cultivation method used allows the Zibibbo grape to grow and vinify, all the same, with lower water volumes. Low-growing vines cultivated in hollows where dew can collect, a very short vine pruning period and a low number of buds per vine enable the plant to survive even under water stress conditions. It is this particular condition (constant water stress) that results in the organoleptic traits much sought after in the Moscato di Pantelleria.

Looking at the water footprints of the analysed wines results that in the north and centre of Italy, where the climate allows for a more limited evapotranspiration, the water footprint is lower than in areas where, instead, there is more sunlight exposure and transpiration.

3.3 DOP Saffron: From San Gimignano, L'Aquila and Sardinia

The saffron selected for this study comes from the plant, *Crocus sativus L.* The production methods for DOP saffron in different areas of Italy are very similar to each other and are governed by product specifications and rules (disciplinare di produzione dello zafferano) for the product's cultivation and production. These specifications and rules, different to the other two product categories studied (Parmigiano Reggiano and Barolo, Sassicaia and Moscato di Pantelleria wines), permit the soil fertilisation. This soil nutrient enrichment involves a prior fertilisation of the saffron (the use of manure). The specifications and rules for Sardinian saffron also permit crop rotation, by means of a catch crop. In fact, grain legumes (beans, chickpeas, etc.) can be planted on land to be used for Sardinian saffron cultivation the year prior to the farming of the spice. This operation enriches the soil with nitrogen by means of nitrogen fixation resulting from microorganisms which live in symbiosis with the roots of this particular type of plant. The farming of catch crops renders the soil more fertile. This practice, only permitted by the *disciplinare* for Sardinian saffron, allows for assessing the water footprint according to two different methods for improving soil fertility—natural fertilisation and crop rotation. Where irrigation is concerned, the product specifications and rules prohibit any irrigation of the saffron plant.

To calculate the water footprint in the case of natural fertilisation, in accordance with the *disciplinare* of Sardinian saffron product, an estimate of the fertilisation of about 300 cwt/ha of manure prior to the beginning of the growth cycle was used. If it is assumed, according with the CRPA guide (Animal production research centre on organic fertiliser use) that the moisture content of the manure is 30 %, it results that fertilisation contributes to 90 cwt/ha of water. Then, referring to the

Table 4 Water footprint of three types of DOP saffron by using manure fertilisation only

	Water footprint (l/gm)
Sardinia (San Gavino)	581
Tuscany (San Gimignano)	761
Abruzzo (L'Aquila)	698

Source edited by the authors

meteorological data of the production areas and the plant's life cycle and productivity, the water footprint for the saffron produced in the three zones was determined (Table 4). The productivity, according with the sector monographs, was placed at 10 kg/ha per year.

The effect of a crop forcing such as manure fertilisation does not greatly change the resulting water footprint. On the contrary, the high water footprint of saffron is highlighted (litres/gram), reaching the same high level as the footprint for vanilla. In fact, to produce 1 kg of saffron stigmas, 200,000 flowers are needed.

4 Scenarios

The hypothesis is raised for a DOP Parmigiano Reggiano produced in a climatic region different to the traditional one (but always plausible from an agronomic viewpoint). Italy with a mainly north–south axis exhibits different climate and subclimate conditions down its axis. To analyse the water footprint's susceptibility to different production sites, cases were analysed hypothesising the production of Parmigiano Reggiano cheese in different Italian geographic situations. As a comparative case study, results obtained from the production of the cheese in the Trapani area of Sicily are shown. As for the water print calculation of Parmigiano Reggiano in its zone of origin, the hypothesis of irrigation on demand is considered.

As far as the wines are concerned, the comparison was made between the three studied wines and the average water footprint of Italian wines (Hoekstra and Chapagain 2007).

The studies on saffron cultivation led to analysing the same type of plant (*Crocus sativus* L.) grown using crop rotation. This resulted in analysing the impact of a non-water external forcing on the water footprint of saffron.

5 Results

5.1 Parmigiano Reggiano

The two scenarios considered to calculate the water footprint of the Parmigiano Reggiano cheese are based on the different footprint of the dairy cow's feed grown

Table 5 Water footprint of the elements of the dairy cow diet in a real (Parma) and in a fictitious scenario (Trapani)

		Dairy cow diet		
		Barley 4.5 kg/day	Maize 7 kg/day	Fodder 11.5 kg/day
Parma	Water footprint $_{green}$ (l)	3.100	2835	4.800
	Water footprint $_{blue}$ (l)	9	735	–
Trapani	Water footprint $_{green}$ (l)	3.800	2074	6.119
	Water footprint $_{blue}$ (l)	9	2709	136

Source edited by the authors

in the Parma area or in Sicily. A water footprint for 100 g of Parmigiano Reggiano is shown (Table 5).

In general, it can be noted that the water footprints of the fictitious scenario are much higher and, particularly, that the blue water footprint values are very different in the two scenarios. This highlights that if the DOP products are removed from their traditional context, they result in a much higher environmental impact—in this case, the blue water footprint highlights a heavy demand for irrigation in the south of Italy to attain the same quantities of fodder. The impact that these crops have on the environment and, especially, on the water resources is more marked in the south compared to the north of Italy.

The impact of water use from a social point of view (Aldaya and Hoekstra 2010) has not been included in this study. In fact, in Sicily, different to the Parma zone, there is a water scarcity and irrigating the crops would make the product even less sustainable environmentally. Consequently, the higher footprint of the animal feed weighs even more heavily in the cultivation sustainability.

5.2 Wine

The comparison of the average water footprint of Italian wines, which is 708 m^3/t where 543 m^3/t is a green water footprint, 47 m^3/t a blue water footprint and 118 m^3/t a grey water footprint (Hoekstra and Chapagain 2007), and those studied shows how wines produced according to *disciplinare* of the product have a lower water footprint. Therefore, from this analysis, it is clear that quality label (DOC and DOCG) wines guarantee not only a high quality, but also environmental sustainability.

Table 6 Estimate of the water footprint for a gram of DOP Sardinian saffron grown with a catch crop rotation

	Saffron water footprint (l/ha)	Legume water footprint (l/ha)	Total water footprint (l/gr)
Sardinia (San Gavino)	5.800.000	1.212.000	701
Tuscany (San Gimignano)	7.610.000	1.100.000	871
Abruzzo (L'Aquila)	6.980.000	950.000	793

Source edited by the authors

5.3 *Saffron*

The scenario concerning saffron provides a comparison of the water footprint for saffron produced from crop rotation methods and of that produced with natural fertilisation (it should be remembered that the *disciplinare* of Sardinia saffron does permit crop rotation, while San Gimignano and L'Aquila saffron are permitted to use only the manure fertilisation method). Therefore, it became necessary to calculate the legume water footprint used to fertilise the soil. This water footprint was calculated using the CROPWAT 8.0 model and then added to the saffron water footprint (Table 6).

As can be seen in Table 6, the saffron footprint increases, on an average, by 16 % with grain legume cultivation compared to that of natural fertilisation. Therefore, we can confirm that also external actions not involving the direct use of water significantly contribute to modifying the product's water footprint. Specifically, in the case of Sardinian saffron, where the product specifications and rules permit this practice, the water footprint increases by more than 20 % when compared to the manure fertilisation. This should not be considered as a proposal to change the product specifications and rules, but as an explanation for the factors influencing and modifying, sometimes quite markedly, the calculation of the product's water footprint. Indeed, for this work, it appears to be important then to highlight what and how much weight do the single contributions of water in calculating an end product's water footprint carry.

6 Conclusions

From the studies presented, it can be confirmed how the environmental sustainability of the agricultural products is linked to their production zone and how the quality label also underlines, apart from the product's quality, its environmental sustainability. The zone of origin's pedologic, weather and agronomic features endow the product with its qualitative uniqueness not calling for, or in some cases only minimally, any improvements, such as supplementary irrigation or fertilisation, which would increase the product's water footprint. The scenario analyses proposed in this study have highlighted how in relocating a product's production from its original zone also results in, other than the obvious qualitative differences closely linked to the zone which confers its particular organoleptic features, differences in its environmental impact. Therefore, it would be desirable, apart from protecting the DOC, DOP or DOCG quality label, to limit as far as possible the growth of the crop to its zone of origin, thus safeguarding the environment and especially the water resources.

The analysis and the considerations on the proposed scenarios raise questions concerning global changes, on their impact on crop localisation and on water resources.

References

Aldaya, M. M., & Hoekstra, A. Y. (2010). The water needed for Italians to eat pasta and pizza. *Agricultural Systems, 103*, 351–360.

Allen, R. G., Pereira, L. S., Raes, D., & Smith, M. (1998). *Crop evapotranspiration—Guidelines for computing crop water requirements*. FAO irrigation and drainage paper 56.

Bianchi, P. G., & Castelli, P. G. C. (2007). *Manuale di agricoltura*, Hoepli.

Chapagain, A. K., & Orr, S. (2009). An improved water footprint methodology linking global consumption to local water resources: A case of Spanish tomatoes. *Journal of Environmental Management, 90*(2), 1219–1228.

Chapagain, A. K., & Hoekstra, A. Y. (2003). Virtual water flows between nations in relation to trade in livestock and livestock products. In *Value of water research report series n. 13*. UNESCO-IHE, www.waterfootprint.org/Reports/Report13.pdf.

Chapagain, A. K., & Hoekstra, A. Y. (2004). Water footprints of nations. In *Value of water research report series n. 16*. UNESCO-IHE, www.waterfootprint.org/Reports/Report16Vol1.pdf.

Chapagain, A.K., Hoekstra, A.Y., Savenije, H.H.G., & Gautam, R. (2006). The water footprint of cotton consumption: An assessment of the impact of worldwide consumption of cotton products on the water resources in the cotton producing countries. *Ecological Economics, 60* (1), 186–203.

Controlled Denominaztion of Origin for "Bolgheri" and "Bolgheri" Sassicaia wines. Ministerial decree on agricultural resources, 14 June 2001.

EC Regulation n. 510/2006 of the Council 20 March 2006. *Elenco delle denominazioni italiane, iscritte nel Registro delle denominazioni di origine protette e delle indicazioni geografiche protette* (updated 11 December 2010).

Ercin, A. E, Aldaya, M. M., & Hoekstra, A. Y. (2009). A pilot in corporate water footprint accounting and impact assessment: The water footprint of a sugar-containing carbonated beverage. In *Value of water research report series*, n. 39, UNESCO-IHE.

Food and Agriculture Organization of the United Nations. (2009). *CROPWAT 8.0 decision support system*.

Gerbens-Leenes W., Hoekstra, A.Y., & van der Meer Theo H. (2009). The water footprint of bioenergy. *Proceedings of the National Academy of Sciences of the United States of America*

Hoekstra, A. Y., Chapagain, A. K., Aldaya, M. M., Mekonnen, M. M. (2009). *Water footprint manual. State of the art 2009*. The Netherlands: Waterfootprint Network.

Hoekstra, A. Y., & Chapagain, A. K. (2007). Water footprints of nations: Water use by people as a function of their consumption pattern. *Water Resources Management, 21*(1), 35–48.

Hoekstra, A.Y., & Chapagain, A.K. (2008). *Globalization of water: Sharing the planet's freshwater resources*. Oxford, UK: Blackwell Publishing.
Hoekstra, A. Y. (2010). The water footprint of animal products. In J. D'Silva, & J. Webster J. (Eds.), *The meat crisis: Developing more sustainable production and consumption, Earthscan* (pp. 22–33). Londra.
Product specifications for Parmigiano Reggiano DOP cheese D.P.C.M. 4.11.1991.
Product Technical Specifications for DOP Zafferano dell'Aquila, Reg. CEE 2081/92.
Product Technical Specifications for DOP Zafferano di San Gimignano Reg. CEE 2081/92.
Product Technical Specifications for DOP Zafferano di Sardegna, Reg. CEE 2081/92.
Soster, M., Cellino, A. (1998). Caratterizzazione delle produzioni vitivinicole dell'area del Barolo: un'esperienza pluriennale. Quaderni Regione Piemonte, n. 13 October 1998, Agricultural Development Services Sector.
Specifications for Barolo D.O.C.G. Recognition of Controlled and Guaranteed Denomination of Origin of Barolo wine, Barolo D.O.C.G. Specification—D.P.R. 1 July 1980.
Specification for Moscato di Pantelleria D.O.C.—D.M. 27 September 2000.
USDA-SCS. (1985). National engineering handbook, section 4—Hydrology.

Calculating the Water Footprint of an Agri-Food Company's Supply Chain: The Mutti Case

Monia Santini and Riccardo Valentini

1 Introduction

As in general for the whole agricultural sector, quantifying the WF is essential to evaluate water saving strategies in the processes involved from tomato cultivation to industrial processing, in order to obtain the final food products (paste, purée, pulp, sauces, peeled, etc.).

It is commonly known that tomato farming in Italy requires enormous amounts of water (from 300–400 to 7000–8000 m^3/ha), whose availability is conditioned by different factors, particularly the geographic location and the climatic/geomorphological conditions of the cultivation site. Given the scarce (and decreasing) availability of "natural" water resources (rainwater) in some areas, intensive irrigation practices exploiting both surface water and groundwater are often necessary.

Consequently, the first fundamental step in calculating the WF is to distinguish between the quantity of water originating "naturally" and directly from rain and what is "artificially" supplied by irrigation. Secondly, the impacts in terms of water resource pollution due to fertiliser and pesticide use must be considered.

Besides the tomato cultivation phase, water consumption and pollution linked to all the tomato processing phases in the plant up to the end-product being deposited in the warehouse, and ready for market distribution, must not be overlooked.

For this reason, calculating the WF for Mutti was based on a well-assessed methodology (Chapagain and Orr 2009) with, however, some elements being

M. Santini (✉) · R. Valentini
Division Impacts on Agriculture, Forests and Ecosystem Services,
Centro Euro-Mediterraneo sui Cambiamenti Climatici, Viterbo, Italy
e-mail: monia.santini@cmcc.it

R. Valentini
Department for Innovation in Biological, Agro-Food and Forestry Systems (DIBAF),
University of Tuscia, Viterbo, Italy

© Springer International Publishing Switzerland 2015

M. Antonelli and F. Greco (eds.), *The Water We Eat*,
Springer Water, DOI 10.1007/978-3-319-16393-2_18

fine-tuned to improve the inclusion of the different processes in the whole chain, and to make the best possible use of the information available.

While the water use linked to the cultivation phase can be reliably reproduced by means of numerical models which simulate the crop growth stages and their water consumption, quantifying other water uses, such as those related to transportation, energy use (both in the fields and in the factory), washing and production of packaging components, required some assumptions and simplifications whose influence was taken into consideration so that results can be exploited in the appropriate manner.

After a brief description of the approach adopted, of the input data required and of the methodological improvements resulting from the detailed information available, including the analysis of the necessary simplifications and inevitable limitations, this chapter presents the results and their analysis in supporting the company's objectives for WF reduction.

2 The WF Calculation Methodology

2.1 Supply Phase

Using as a reference the works of Chapagain and Orr (2009) and Aldaya and Hoekstra (2010), the WF (expressed in m^3/t) of the tomato production during the cultivation phase (making part of the supply phase—SP) is the ratio between the volume of water used for the crop growth per hectare (used water, UW, m^3/ha) and the yield (crop yield, Y, t/ha):

$$WF = \frac{UW}{Y}$$

where

$$UW = UW_{gn} + UW_{bl} + UW_{gy}$$

with

UW_{gn} = volume of water used for evapotranspiration from rainfall (green water);
UW_{bl} = volume of water used for evapotranspiration from irrigation (blue water);
UW_{gy} = volume of unusable water as polluted by fertilisers and pesticides (grey water).

2.1.1 Green Water Use

To quantify the evapotranspiration, the CROPWAT model (http://www.fao.org/nr/water/infores_databases_cropwat.html), developed and distributed by the FAO (1992), was used. It requires input data regarding weather, pedological characteristics and crop growth parameters in order to simulate the water requirements during the crop development phases.

The water demand for the evapotranspiration for time t during the crop growth period is calculated as follows:

$$ET_C[t] = K_C[t] \times ET_0[t]$$

The ET_0 (reference evapotranspiration, mm/day) depends exclusively on climate, representing the atmospheric evaporative capacity at a given site and in a given period of the year. The plant's evapotranspiration requirement ($\mathrm{ET_C}$, mm/day) differs from the reference evapotranspiration by a factor K_C, which depends on the growth stage (varying from sowing to harvest), on the crop variety and the climate settings.

Instead, the quantity of water available as soil moisture depends on the effective rainfall (R_{eff}), calculated according to different procedures, and on the irrigation I_R.

Therefore, the use of green water for time t ($u_{\mathrm{gn}}[t]$) is obtained from the minimum value between R_{eff} and $\mathrm{ET_C}$.

$$u_{\mathrm{gn}}[t] = \min(R_{\mathrm{eff}}[t], ET_C[t])$$

The total use of green water during the cultivation phase is obtained from the integral, over a daily time step, for the whole growth period (L days), from sowing to harvest.

$$UW_{\mathrm{gn}} = \sum_{t=1}^{L} u_{\mathrm{gn}}[t]$$

2.1.2 Blue Water Use

While the use of green water depends solely on the R_{eff} and $\mathrm{ET_C}$, the use of blue water depends on the $\mathrm{ET_C}$, on green water availability and on the water available from irrigation.

The water requirement from irrigation I_R for time t is calculated as follows:

$$I_R[t] = ET_C[t] - u_{\mathrm{gn}}[t]$$

The use of blue water is obtained from the minimum value between the irrigation demand $I_R[t]$ and the water effectively supplied by irrigation $I_{\mathrm{eff}}[t]$ that is part of the

$I_R[t]$ and depends on the soil conditions. The use of blue water decreases to zero if the total crop water requirement for evapotranspiration is satisfied by R_{eff}.

$$u_{\text{bl}}[t] = \min(I_R[t], I_{\text{eff}}[t])$$

As for the green water, the total use of blue water is therefore calculated integrating:

$$UW_{\text{bl}} = \sum_{t=1}^{L} u_{\text{bl}}[t]$$

2.1.3 Grey Water Use

The volume of dilution water (grey element) is the theoretical quantity of water required to dilute the pollutants (generally, compounds of nitrogen, phosphorus, potassium and pesticides) used during the cultivation phase, so that the water quality in soils and water bodies remains above the legal qualitative standards.

The use of grey water for time t is obtained from

$$u_{\text{gy}}[t] = \max\left(\frac{Q[i,t]}{Q_{\max}[i,t]}\right)$$

where $Q[i, t]$ is the quantity (ton) of pollutant i introduced during cultivation time $[t]$, while $Q_{\max}[i, t]$ (t/m^3) is the quantitative maximum permitted in underground and/or surface water bodies.

The total volume of grey water use (for dilution) can be, therefore, calculated as follows:

$$UW_{\text{gy}} = \sum_{t=1}^{L} u_{\text{gy}}[t]$$

2.2 *Operative Phase*

Following the cultivation phase, in order to calculate the WF of the operative phase (OP), each processing step must be clearly spelt out. Moreover, it must be considered that the same input product (in this case, the harvested tomato) results in N end products (paste, purée, pulp, etc.).

Therefore, initially, to calculate the WF for each output product n (WF_n), a proportion of the WF used in the previous step of the SP must be added (Wf_{in}) to the

WF of the processing steps (WFp_n); this proportion is based on the ration (fp_n) between the quantity of output product n and the respective quantity of input product needed for obtaining n.

The final WF (WF_{FIN}) is calculated from the sum of the WF_n of the N output products.

$$\mathrm{WF_{FIN}} = \sum_{n=1}^{N} (\mathrm{WF_n}) = \sum_{n=1}^{N} \left(\frac{\mathrm{WF_{in}}}{\mathrm{fp_n}} + \mathrm{WFp_n} \right)$$

3 The Mutti Case: The Scope of Application and Data

Based on the approach used to calculate the WF described in the previous paragraph, following the collection of the necessary appropriate data, when easily accessible and directly available, the next steps involved the analysis, selection and implementation of procedures for correcting the primary data and processing them to derive further secondary data.

To define the spatiotemporal domain for the application of the WF calculation procedure, the right compromise was sought between the speed and cost of the retrieval of useful data and their spatiotemporal detail/coverage.

Accordingly, to assess the WF for the year 2010, for the SP, the farms representing 47 % of the tomatoes supplied to Mutti were considered, aggregating their data at municipality level, which became the spatial unit of the analysis.

After their collection, processing and standardization, data were organised into a geodatabase, i.e. an explicitly referenced geographic base of data.

3.1 Data to Calculate the WF in the Supply Phase (SP)

3.1.1 Ingredient Data and Parameters

A number of farm companies (each responsible for one or more fields) which, in turn, are members of farms' associations, provide tomatoes to the Mutti factory.

In 2010, Mutti bought tomatoes from 13 farm's associations plus a small part from the open market. Considering that the Interprovincial Association of Fruit and Vegetable producers, AInPO, is Mutti's main supplier, it was chosen to calculate the WF based on its cultivation data and, then, to rescale the final WF based on the proportion between the AInPO and total supply.

The data sheets for each AInPO member (farm company), digitized and stored at the Emilia Romagna Region Agricultural Department office, were analysed. They describe the agricultural practices carried out during the tomato cultivation phase and for the various fields managed by each company. The information mainly concerns the sowing and harvesting dates and the irrigation, fertiliser and pesticide applications (timing and quantity).

The other ingredient in tomato-based food products which was taken into account was salt; in this case, Mutti supplied the figures for the salt used in 2010.

3.1.2 Meteorological Data

The next step conducted was the collection of information on climate variables' time series required for the studied period. To guarantee a thorough coverage of all the phases of sowing, fertilisation, irrigation, plant health controls and harvesting, the reference year for the search data was from 1 November 2009 to 31 October 2010.

The monthly weather variables required for the CROPWAT model are as follows: average minimum and maximum temperatures; average air relative humidity; average wind speed; average hours of sunshine; and rainfall amount.

Given the municipalities are all situated in three regions—Piedmont, Lombardy and Emilia Romagna—the research first aimed at collecting the time series on the above variables for the meteorological stations managed by territorial and/or environmental agencies (ARPA Piedmont, ARPA Lombardy and ARPA Emilia Romagna). The station data were spatially averaged to obtain the representative values for each municipality where the cultivation was to be studied.

3.1.3 Soil Data and Parameters

In order to obtain, for each municipality, the soil parameters required to apply the CROPWAT model, the most up to date, homogeneous data set was used—the Harmonized Soil World Database (HSWD, http://www.iiasa.ac.at/Research/LUC/External-World-soil-database/HTML/)—providing the highest spatial resolution available (1 km) and including the required soil attributes: texture in terms of sand, silt and clay percentages which, based on the dominant fraction, allow classifying soils as sandy, silty or clayey.

3.2 Data to Calculate the Secondary and Indirect WF

In the SP, besides the primary product (tomato) cultivation, a component of the WF (defined as "secondary") derives from the production of components such as containers (cans, tubes, bottles) and their accessories (lids, labels), packaging (trays, stretch film wrapping) and equipment used in the factory (pallets). Obviously, the spatiotemporal detail of the information required to quantify these added components is totally different from the data described in the previous paragraphs, as aggregate estimates are available for the single production year.

Unfortunately, no established models or standardised procedures exist to measure this element of the WF. However, an estimate was possible thanks to data on the quantity of material for the above-listed typologies being provided by Mutti for the 2010 campaign and using the literature methods (Ercin et al. 2009; Botto 2009; Li and Nwokoli 2010). It was also assumed, based on the information gathered, that no recycled elements were used.

The water volume used in the SP regarding transportation, energy, etc. (indirect component), was estimated as a percentage of the water required in the primary SP according to Ercin et al. (2009).

3.3 Data to Calculate the WF in the Operative Phase (OP)

After the quality control carried out on a primary product (tomato) sample delivered by the supplier companies to Mutti, the tomato is ready for the following processing phases, described in Table 1 for the various product categories.

From the study of these procedures, the following components to calculate the operative phase WF were considered:

- water consumed and/or polluted during the production processes (washing, bottling, labelling, packaging, etc.);
- water consumed and/or polluted by workers for domestic use (drinking, sanitation).

For these two components, Mutti provided an estimate for the 2010 internal water accounting, with the water volume aggregated by water consumption typology.

Table 1 Tomato processing steps following washing and sorting

Product category	Process
Pulp	Dicing, draining, filling, pasteurising, canning/bottling
Pureed	Hot break dicing, refining, pureeing, pasteurising, canning/bottling
Paste	Hot break dicing, refining, pureeing, pasteurising, canning/tubing
Juices	Hot break dicing, refining, pureeing, pasteurising, canning/bottling
Ready-made sauces (vegetable pastes/sauces)	Formulating, pasteurising, packaging
Ketchup	Formulating, pasteurising, packaging
Pizza sauce	Formulating, pasteurising, packaging. An extension of the "pureed" production cycle
Vinegar	Filtering, pureeing, fermenting, bottling

Source edited by the authors

As no rainwater is used in the OP, the green water element for this phase was considered to be zero.

Furthermore, all the water consumed, before reaching the final wastewater discharge, is treated. The concentrations of the same elements considered for the fertilisers (N, K, P), analysed after the treatment, were found to be greatly lower than the maximum allowable concentrations and, consequently, it was assumed that the OP grey element was also zero.

Given these assumptions, from the annual water accounting table provided by Mutti, the blue element was calculated and subdivided into domestic discharge, wastewater and loss due to evaporation:

- 10,500 m^3/year of domestic discharge;
- 512,000 m^3/year of wastewater, from the difference between the 622,000 m^3/year produced and the 110,000 m^3/year recycled (17.7 % recycle index);
- loss due to evaporation, corresponding to 50,000 m^3/year;
- drinking water use, corresponding to 36.6 m^3/year.

4 Implementation of the Methodology

The information available for the AInPO suppliers was spatially aggregated for the 39 municipalities involved in the study (as farms are spatially distributed across these municipalities). The number of farms was reduced aggregating, within the same municipality, those with the same sowing and harvesting dates, so to assume the same growth cycle to be reproduced by CROPWAT model to calculate the water required in the various phases (subdivided into 10 day periods). The total number of simulated groups of farms' fields was 583.

Moreover, to take into account the uncertainty derived from different available methods to simulate the effective rainfall R_{eff}, three different procedures were applied:

- the R_{eff} is considered a fixed percentage (80 %) of total rainfall;
- the model conducts statistical analyses (method developed by the FAO), starting from monthly rainfall data provided, of the probability that a given effective rainfall threshold is exceeded for each 10 day period used in the temporal simulation of the growth phases;
- an empirical formula developed by the US Department of Agriculture (USDA) is used.

Therefore, the three procedures to calculate the R_{eff} were applied via CROPWAT in all the 583 fields' groups, with a total of 1749 simulations conducted. Finally, the three quantitative results for the green and blue water, derived from using the three different formulae to calculate the R_{eff}, were averaged for the same group of fields.

Instead, where irrigation is concerned, despite having information on the methods used such as sprinkler irrigation and micro-irrigation (by far the most used), an average efficiency of 70 % was assumed.

Moreover, it should be noted that the blue element was supplemented by the component due to another ingredient—salt—calculated by assuming that 1 kg of salt may be extracted from 30 l of seawater. Consequently, the virtual water contained in the reported quantities of salt corresponds to about 14,625 t.

Instead, more simplified models were used to quantify the fertilisers and pesticides leached through the soil and hypothetically entering the surface/underground water bodies.

As far as the chemical fertilisers (combination of N, K, P) are concerned, the nitrogen and potassium leaching were based on the approach in Smaling et al. (1993), while for phosphorus, the leaching module in the SWAT hydrological model (Arnold et al. 1998) was used.

Although this approach is simplified as it considers that nutrient release and leaching occur at the same rate, and it uses annual rainfalls rather than their distribution (Elias et al. 1998), it is undoubtedly a big step forward compared to previous approaches where an approximate leaching of 10 % (and only for nitrogen) was considered.

For pesticides, given the lack of literature available, a leaching factor of 0.5 % was used, obtained from an average for the silt/sandy soils and for rainfall patterns similar to those of the studied area (Petrovic 1995).

In the calculation steps, taking into account the most recent guidelines and regulations, maximum concentration limits were considered as follows:

- for N and P, 15 and 2 mg/l, respectively, both representing the maximum allowable for discharge into the soil—see Table 4, Annex 5 in the Italian Dgls 152/99);
- for K, 10 mg/l (considering the only, and quite precautionary, standard available as it refers to drinking water use http://www.dairyforall.com/dairy-waterquality.php).

Concerning pesticides, the maximum value considered allowable is 0.05 mg/l (discharge into surface waters—see Table 3, Annex 5 of the Italian Dgls 152/99).

5 Results and Discussion

This paragraph presents the results for the different phases and components (primary, secondary and indirect SP, and OP), and the WF elements (green, blue, grey).

The WF for the primary SP results as follows:

- green: 49.03 m^3/t;
- blue: 114.53 m^3/t;
- grey: 161.67 m^3/t;
- total: 325.23 m^3/t.

It can be clearly seen that the grey element is responsible for practically half of the WF.

Two factors were taken into account for a correct quantification of the primary SP component:

- not all the tomatoes grown by the suppliers were delivered to Mutti. Consequently, the WF was calculated using the average yield data (t/ha) from AInPO;
- the AInPO association accounts for about 47 % of the tomatoes delivered to Mutti, and therefore, the cultivation phase WF (m^3/t) was multiplied by the total (in tonnes) delivered to the company so to derive the actual cubic metres of water consumption corresponding to the cultivation phase for tomatoes allocated to Mutti.

Instead, in the case of the SP's "secondary" step (packing/packaging material production), it can be clearly seen that the grey element (30.39 m^3/t) accounts for almost 70 %, due to the enormous use of paper and cardboard, while, compared to the previous result, the green element (6.43 m^3/t) is much lower, almost equal to the blue (6.61 m^3/t).

Where the "indirect" phase (use of energy and transportation) is concerned, the WF values are 0, 0.03 and 16.36 m^3/t for the green, blue and grey elements, respectively, demonstrating how the water pollution element (99.8 %) is practically the only influencing factor.

Summing up the SP's primary, secondary and indirect WF, the following results were obtained:

- green: 55.46 m^3/t;
- blue: 121.17 m^3/t;
- grey: 208.42 m^3/t;
- total: 385.05 m^3/t.

Analysing the weight of each WF element in the different components (primary, secondary and indirect), we see that:

- for green WF, the production of ingredients account for 88.2 % and packaging (mainly the use of paper/cardboard) for 11.8 %;
- for blue WF, the production of ingredients account for 94.5 % and packaging (mainly the use of paper/cardboard) for 5.5 %;
- for grey WF, the production of ingredients account for 77.55 %, packaging (mainly the use of aluminium/glass/PET) for 14.6 % and energy use 7.85 %.

It is clear that the cultivation phase has the highest impacts with its three elements, playing a not so negligible role (well over 10 %) in the "secondary" phase for the green and grey WF.

The three-dimensional graph in Fig. 1 summarises the above, confirming how the grey element and the primary component play a key role in the WF for all the SP.

Considering that the annual water use for the processing phases in the factory is 572,537 m^3, the OP blue element (the only one) is 7.87 m^3/t.

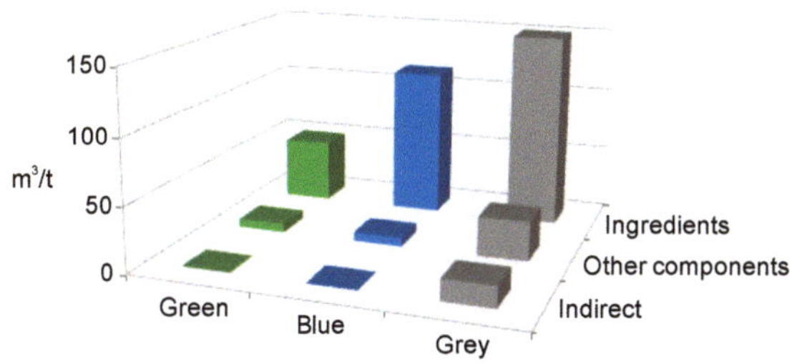

Fig. 1 The impact of the different components and elements of the WF total for the SP. *Source* edited by the authors

Summing up the components of the supply and operative phases (SP and OP), the following WF values can be obtained:

- green: 55.46 m^3/t (14 %);
- blue: 129.03 m^3/t (33 %);
- grey: 208.42 m^3/t (53 %);
- total: 392.92 m^3/t.

Given the minor role (approximately 2 %) of the OP component, this result confirms once again how the SP grey and blue elements, affected more by human activity rather than natural conditions, account for more than a half and a third, respectively, of the total.

Even with the inevitable simplifications and assumptions, the methodology implemented provides, besides an estimate representative of Mutti's present situation, all the tools required to check and repeat the calculations for future years and in different localities where cultivation field should be situated.

6 WF Reduction Objectives and Proposed Strategies

Differently than for the carbon footprint, there is no literature available about the WF that would allow for individuating benchmarks regarding reduction objectives to be pursued and strategies to be adopted for successful objective achievement.

Nevertheless, the pioneering work done allows us to make general and preliminary evaluations, which can be verified and reviewed, and can pave the way for a commitment to a WF reduction.

From the results, it emerges that the WF can be mainly attributed to the SP (especially, the tomato cultivation). The fact that Mutti does not directly manage the tomato cultivation fields means that it has no direct control over this phase's operations. However, awareness-raising campaigns aimed at the farmers could be promoted concerning those practices resulting in a high water consumption impact (pesticide and fertiliser use and irrigation techniques).

As well, we can see how the SP's secondary step plays quite an important role and, therefore, the objectives of WF reduction or, however, resource saving, could also involve company choices regarding the products used in its factory for the packaging/packing phases.

In the following, a preliminary feasibility study is shown for those strategies which could potentially result in a significant WF reduction and, therefore, support a reasonable and realistic commitment to a WF reduction.

6.1 Reduction Hypotheses' Impact on the WF and Feasibility Study

In order to carry out an initial rough estimate about the reduction potential of supply chain processes and relying on the results attained, the impact that the following four hypotheses can have on the WF was assessed:

1. improvement in irrigation method efficiency (e.g. micro-irrigation rather than the sprinkler system, possibly increasing efficiency from 70 to 90 %);
2. reduction in pesticide use of 50 %;
3. reduction in fertiliser use of 50 %;
4. reduction in pesticide and fertiliser use of 25 % each.

Table 2 shows the potential WF reduction percentages in the case of the four hypotheses presented, for each element and the total.

Such analysis is not so much aimed at defining the exact reduction figures, but more at supporting an effective action to raise farmers' awareness of reducing the impact of their cultivation practices on the water resources, without negatively influencing the product quality/quantity.

Therefore, a more in-depth analysis of the potential and effective applicability of the hypothesised strategies was carried out, based on the tools used by the farmers in the studied area.

Two of the Mutti suppliers conducted experiments in the 2011 season and their preliminary results have recently been examined.

(a) The first experiment consisted in comparing the irrigation water consumption between two farms where, in one case (traditional), the water requirements

Table 2 WF reduction percentages estimated for each of the 4 hypotheses described in the text

Hypotheses	Water footprint reduction %			
	Green	Blue	Grey	Total
1	0	19.69	0.00	−6.47
2	0.00	0.00	−32.89	−17.45
3	0.00	0.00	−9.82	−5.21
4	0.00	0.00	−21.36	−11.33

Source edited by the authors

were estimated based on the regional weather bulletins, and in the other case (innovative), by making use of probes with sensors (tensiometers) planted in the ground to monitor in nearly real-time the soil moisture at different depths, allowing to regulate ad hoc irrigation applications according to monitored soil water needs. The preliminary findings showed that the innovative method resulted in reducing the number of irrigations required and lowering the quantity of water provided for each irrigation turn, leading to overall water saving of at least 10 %.

(b) The second experiment consisted in comparing farms using traditional fertilisation methods and others using ferti-irrigation which avoids an excessive loss in the nutrient elements. The findings showed that, even by reducing the N and K application per hectare (−31.4 and −32.1 %, respectively) and increasing the P application (+7.8 %) thanks to the innovative method, the resulted yield increased, however, of more than 8.5 %.

Based on these two results, with a 10 % reduction in irrigation and an average 25 % reduction in fertilisers, the WF calculation was repeated assuming a uniform adaptation of strategies (a) and (b) in all the supplier farms obtaining, in the first case, a potential reduction of 2.65 % and in the second of 2.52 %. Instead, considering a combination of the two innovative methods (a) and (b), a potential reduction of 5.16 % was estimated.

In the successive years, further experiments were set up in this direction confirming the validity of WF reduction strategies identified.

7 Conclusions

This study has highlighted how the agricultural practices adopted during the cultivation phase have a strong influence on the WF of agri-food products.

In the absence of any benchmark to quantify the reduction target and analysing the tools already available to the local farmers, it becomes immediately evident that, firstly, the reduction objectives should depend on the raising awareness of the farmers being an active part of the production chain, by promoting the monitoring of the water, fertiliser and/or pesticide requirements and optimising applications. The use of tools for the control of soil moisture and fertility levels (as well as pest prevention and controls) can support a more effective and efficient planning for the different practices, resulting in a both quantitative (water saving) and qualitative (pollution reduction) improvement.

The two experiments carried out to investigate WF reduction potential of farming practices confirmed that improving irrigation and fertilisation techniques (and so their efficiency) results in advantages in terms of water saving. Thus, the Mutti's priority in a water reduction commitment should be raising farmer awareness about the importance of adopting monitoring/alert tools regarding soil moisture and fertility conditions and, consequently, useful in scheduling and optimising cultivation practices.

Moreover, it was noted that if combining multiple promising strategies, more significant reduction objectives can be achieved (and/or in less time).

Based on these findings, Mutti, in collaboration with the WWF, has embarked on a WF reduction programme using measures to improve efficiency and effectiveness in irrigation methods and fertiliser reduction. It is an innovative project that has led to identifying more efficient solutions, actively involving the supply chain thanks to information campaigns and company investments in technologies to reduce water consumption.

The Mutti experience shows how by means of these actions it is possible to reduce the impact in an important sector, such as agriculture, bringing the supply chain more in line with measurable and tangible environmental objectives.

Acknowledgments Thanks to the staff of Mutti SpA and WWF Italy for their cooperation in this work.

Bibliography

Aldaya, M. M., & Hoekstra, A. Y. (2010). The water needed for Italians to eat pasta and pizza. *Agricultural Systems, 103*(6), 351–360.

Arnold, J. G., Srinivasan, R., Muttiah, R. S., & Williams, J. R. (1998). Large area hydrologic modeling and assessment—Part I: Model development. *Journal of the American Water Resources Association, 34*(1), 73–89.

Botto, S. (2009). Tap Water vs. Bottled Water in a Footprint Integrated Approach. *Nature Precedings* http://precedings.nature.com/documents/3407/version/1.

Chapagain, A. K., & Orr, S. (2009). An improved water footprint methodology linking global consumption to local water resources: A case of Spanish tomatoes. *Journal of Environmental Management, 90*(2), 1219–1228.

Elias, E., Morse, S., & Belshaw, D. G. R. (1998). Nitrogen and phosphorus balances of Kindo Koisha. farms in southern Ethiopia. *Agriculture, Ecosystems & Environment, 71*(1–3), 93–113.

Ercin, A. E., Aldaya, M. M. & Hoekstra, A. Y. (2009). *A Pilot in Corporate Water Footprint Accounting and Impact Assessment: The Water Footprint of a Sugar-Containing Carbonated Beverage*. Value of water research report series no. 39, UNESCO-IHE.

FAO. (1992). Martin Smith (ed.), *CROPWAT: A Computer Program for Irrigation Planning and Management*, Irrigation and Drainage Paper 46, Roma.

Flörke, M., & Alcamo, J. (2004). *European outlook on water use* (p. 83). Kassel: Final report, Center for Environmental Systems Research.

Li, C., & Nwokoli, S. U. (2010). Investigating the water footprint of Tetra Pak Carton Economy's Beverage Portfolio. *Vatten, 66*, 113–124.

Petrovic, A. M. (1995). The impact of soil type and precipitation on pesticide and nutrient leaching from fairway turf. *USGA Green Section Record, 33*(1), 38–41.

Smaling, E. M. A., Stoorvogel, J. J., & Windmeijer, P. N. (1993). Calculating soil nutrient balance in Africa at different scales. *Fertilizer Research, 35*(3), 237–250.

MIX
Papier aus verantwortungsvollen Quellen
Paper from responsible sources
FSC® C105338

If you have any concerns about our products,
you can contact us on
ProductSafety@springernature.com

In case Publisher is established outside the EU,
the EU authorized representative is:
Springer Nature Customer Service Center GmbH
Europaplatz 3, 69115 Heidelberg, Germany

Printed by Libri Plureos GmbH
in Hamburg, Germany